Studies in Fuzziness and Soft Computing

Volume 442

The series "Studies in Fuzziness and Soft Computing" contains publications on various topics in the area of soft computing, which include fuzzy sets, rough sets, neural networks, evolutionary computation, probabilistic and evidential reasoning, multi-valued logic, and related fields. The publications within "Studies in Fuzziness and Soft Computing" are primarily monographs and edited volumes. They cover significant recent developments in the field, both of a foundational and applicable character. An important feature of the series is its short publication time and world-wide distribution. This permits a rapid and broad dissemination of research results.

Indexed by SCOPUS, DBLP, WTI Frankfurt eG, zbMATH, SCImago.

All books published in the series are submitted for consideration in Web of Science.

Zakiah Mohd Yusoff · Zuraida Muhammad · Siti Aminah Nordin

Fuzzy Logic, Smart Control, and the Future of Essential Oils

The Science of Aroma

Springer

Zakiah Mohd Yusoff
Faculty of Electrical Engineering
Universiti Teknologi MARA (UiTM)
Shah Alam, Selangor, Malaysia

Zuraida Muhammad
Faculty of Electrical Engineering
Universiti Teknologi MARA (UiTM)
Permatang Pauh, Pulau Pinang, Malaysia

Siti Aminah Nordin
Faculty of Electrical Engineering
Universiti Teknologi MARA (UiTM)
Pasir Gudang, Johor, Malaysia

ISSN 1434-9922 ISSN 1860-0808 (electronic)
Studies in Fuzziness and Soft Computing
ISBN 978-981-92-2083-0 ISBN 978-981-92-2084-7 (eBook)
https://doi.org/10.1007/978-981-92-2084-7

This Springer imprint is published by the registered company Springer Nature Singapore Pte Ltd.
The registered company address is: 152 Beach Road, #21-01/04 Gateway East, Singapore 189721, Singapore

Preface

Imagine peeling a fresh kaffir lime and being instantly hit by its sharp, zesty aroma. That fragrance is carried by the essential oils hidden inside the peel, tiny drops of nature's chemistry that have been prized for centuries in food, fragrance, and medicine. But here's the catch: the way we extract those precious oils hasn't really changed much in decades. Steam distillation still rules, and while it works, it often wastes energy, gives inconsistent results, and sometimes destroys the very compounds we want to preserve.

That's where this book begins—with a simple but powerful idea: *what if we could make essential oil extraction smarter?*

This book is the story of how engineering and chemistry can shake hands to solve a real-world problem. It's about taking tools like fuzzy logic, self-tuning control systems, and real-time monitoring usually seen in high-tech industries and applying them to something as natural and timeless as essential oils. Along the way, we discovered that controlling a process isn't just about faster response times or lower error values. It's about aroma, quality, sustainability, and the trust of consumers who want products that are both authentic and consistent.

We wrote this book for anyone curious about how science and technology can transform traditional practices. If you're a researcher or student, you'll find insights into modeling, simulation, and intelligent controllers. If you're from the essential oil industry, you'll see how modern control can protect both yield and quality, giving your products an edge in the market. And if you're simply fascinated by the idea of blending engineering with nature, this book will show you just how powerful that marriage can be.

Throughout the chapters, you'll meet the "characters" of this story: controllers like PID, Hybrid Fuzzy-PID, and Self-Tuning Fuzzy-PID each tested under different conditions, each with strengths and weaknesses. You'll also see how small changes, like adjusting temperature with precision, can make the difference between a vibrant, high-quality oil and one that falls flat.

For us, writing this book has been more than documenting results; it has been about sharing a journey of curiosity, experimentation, and discovery. It has shown us that research can be exciting, practical, and deeply connected to everyday life.

Our hope is that as you read, you'll not only understand the findings but also feel the excitement of asking bold questions and finding meaningful answers.

So, take a deep breath, imagine the scent of fresh lime in the air, and join us on this journey. Let's explore how intelligent control can unlock the full potential of nature's hidden treasures.

Shah Alam, Malaysia — Zakiah Mohd Yusoff
Permatang Pauh, Malaysia — Zuraida Muhammad
Pasir Gudang, Malaysia — Siti Aminah Nordin

Acknowledgements Alhamdulillah, all praise to Allah SWT for granting us strength and perseverance throughout this journey.

Our deepest gratitude goes to our supervisors, Prof. Madya, Dr. Mohd Hezri Fazalul Rahiman and Prof. Ir. Ts. Dr. Mohd Nasir Taib, whose guidance, wisdom, and encouragement have been the backbone of this work. Also, thanks to the Faculty of Electrical Engineering, Universiti Teknologi MARA (UiTM).

To our beloved families, thank you for your endless love, patience, and sacrifices. You are the reason we strive and succeed.

To our colleagues, students, and friends—your support and inspiration have left a lasting mark on this journey.

This book is not just written by us; it carries the spirit of everyone who has walked beside us.

Competing Interests The authors have no competing interests to declare that are relevant to the content of this manuscript.

Blurb

What if technology could unlock the very soul of nature?

For centuries, essential oils have carried the power of fragrance, healing, and tradition. Yet the way we extract them has remained stubbornly old-fashioned, wasting energy, losing quality, and hiding the full potential of every drop.

This book takes you into the heart of a transformation. By blending fuzzy logic, intelligent controllers, and real-time process monitoring, the authors show how modern engineering can revolutionize essential oil extraction. It is a journey from simulation to real-world application, where science meets aroma, and numbers meet nature.

But *The Science of Aroma* is more than a technical breakthrough, it is a vision for the future. A future where extraction is smarter, greener, and more sustainable. A future where communities and industries alike benefit from higher quality, consistent yield, and consumer trust. A future where science and technology empower society and culture.

Written with clarity and passion, this book speaks to researchers, students, industry leaders, and anyone curious about how intelligent systems can breathe new life into ancient traditions.

★ Step inside, take a deep breath, and discover how the fusion of engineering and chemistry is shaping the future of essential oils!

Contents

About the Authors

Assoc. Prof. Ts. Dr. Zakiah Mohd Yusoff is Associate Professor who is currently working at UiTM Shah Alam, Selangor, Malaysia. She received the B.E. in Electrical Engineering and Ph.D. In Electrical Engineering from UiTM Shah Alam, in 2009 and 2014, respectively. In May 2014, she joined UiTM Pasir Gudang as a teaching staff. Her major interests include process control, machine learning, system identification, internet of thing (IoT), and essential oil extraction system. e-mail: zakiah9018@uitm.edu.my

Dr. Zuraida Muhammad is a senior lecturer who is currently working at UITM Cawangan Pulau Pinang. She received the B.E. in Electrical Engineering and Ph.D. in Electrical Engineering from UITM Shah Alam, in 2009 and 2015, respectively. In February 2015, she joined UITM Pulau Pinang as a teaching staff. Her research interests include process control, system identification, big data and essential oil extraction systems. e-mail: zuraida321@uitm.edu.my

Ts. Dr. Siti Aminah Nordin is a distinguished senior lecturer currently affiliated with UiTM Pasir Gudang. She earned both her master's and Ph.D. degrees in Electrical Engineering from UiTM Shah Alam in 2014 and 2022, respectively, showcasing her commitment to academic excellence. With a specialized focus in the realm of Electrical Engineering, her research interests are notably centered around microwave filters, antennas, and electromagnetic wave area. Her academic journey and expertise reflect a deep dedication to advancing knowledge and contributing to the field, particularly in areas crucial to modern communication and technology.
e-mail: sitia181@uitm.edu.my

Symbols and Abbreviations

°C	Degree Celsius
u_0	Chosen amplitude
t	Time
τ	Time constant
K	Process gain
θ	Dead time
V	Voltage
N	Interval
N_s	Switching sequence
T_{sw}	Clock period
Δt	Sampling time
$u(t)$	Input
$y(t)$	Output
$e(t)$	Error
$A(q)$	Output Polynomial
$B(q)$	Input Polynomial
q^{-1}	Backward shift operator
$a_1 \ldots a_n$ and $b_1 \ldots b_n$	Output and input parameters to be estimated
$\varphi(t)$	Regression vector
$\hat{y}$	Predicted output
ε	Residuals
$\hat{\varepsilon}(t)$	Error between desired and actual output
$\bar{y}$	Average for actual output
$\bar{\varepsilon}$	Average residuals
$\bar{u}$	Average input
Kp	Proportional gain
Ki	Integral gain
Kd	Derivative gain
K_C	Controller gain
T_i	Integral time
T_d	Derivative time

M^A and M^B	Fuzzy sets
χ	Universe of discourse
r(t)	Reference input
$\tilde{P}$	Linguistic value
$\tilde{Q}$	Consequent of the rule
Δe	Derivative error
$\%OS$	Percentage overshoots
ζ	Damping ratio
t_r	Rise time
ω_n	Natural frequency
t_s	Settling time
α_0	Ractional change in resistance per degree of temperature at T_o
U_i	Centre of the output membership function
$\mu(y)$	Degree of fulfillment
T_c	Time constant
T_d	Time delay
N_B	Clock period factor
tf	Transfer functions
1-SAP	One-Step-Ahead Prediction
ACF	Auto-Correlation Function
AI	Analogue Input
AIC	Akaike's Information Criterion
ANN	Artificial Neural Networks
AO	Analogue Output
ARMAX	Auto Regressive Moving Average with Exogenous Input
ARX	Auto-Regressive with Exogenous Input
ASTM	American Society for Testing and Materials
BJ	Box-Jenkins
BS	British Standards
CA	Center Average
CCF	Cross-Covariance Function
CO_2	Carbon dioxide
COG	Center of Gravity
D	Derivative
DAQ	Data Acquisition
DCS	Distributed Control System Laboratory
EO	Essential Oil
EOA	Essential Oil Association of USA
FIR	Finite Impulse Response
FLC	Fuzzy Logic Control
FPE	Final Prediction Error
GC	Gas Chromatography
GC-FID	Gas Chromatography–Flame Ionization Detector
GC-MS	Gas Chromatography–Mass Spectrometry
HFPID	Hybrid Fuzzy PID

I	Integral
ISO	International Organization for Standardization
MDL	Minimum Description Length
MF	Membership Function
MSE	Mean-Square-Error
NARMA	Nonlinear Auto Regressive Moving Average
NARMAX	Nonlinear Auto Regressive Moving Average model with Exogenous inputs
NARX	Nonlinear Auto Regressive with Exogenous input
NNARX	Neural network Auto Regressive with Exogenous input
NSSE	Normalized Sum of Squared Errors
OE	Output-Error
P	Proportional
PEM	Prediction Error Method
PI	Proportional-Integral
PID	Proportional-Integral-Derivative
PLCF	Polynomial Linear Curve Fitting
PRBS	Pseudo Random Binary Sequence
R^2	Coefficient of determination
RBS	Random Binary Sequence
RI	Refractive Index
RMSE	Root Mean-Square-Error
RTW	Real-Time Workshop
SCO_2	Supercritical Carbon dioxide
SFE	Supercritical Fluid Extraction
STFPID	Self-Tuning Fuzzy PID
SWE	Sub Critical Water Extraction
TSE	Turkish Standards Institute
V	Voltage

List of Figures

List of Tables

Chapter 1
Essential Oils: Extraction Methods and Challenges

Abstract Essential oils are valuable natural products widely used in pharmaceutical, food, cosmetic, and wellness industries due to their therapeutic and functional properties. Steam distillation remains the most common extraction method because of its simplicity and scalability; however, conventional processes face significant challenges, including long extraction time, low efficiency, loss of volatile compounds, and thermal degradation caused by poor temperature regulation. These limitations directly affect both yield and product quality. Various alternative techniques, such as hydro distillation, solvent extraction, Soxhlet extraction, supercritical fluid extraction, and ultrasound-assisted methods, have been developed to improve performance. Nevertheless, temperature control remains the most critical factor influencing extraction efficiency and preservation of thermolabile compounds. Conventional control approaches, including manual operation, on–off control, and PID controllers, often struggle to manage the nonlinear and dynamic behavior of extraction systems. To overcome these limitations, intelligent control strategies such as fuzzy logic, neural networks, and hybrid fuzzy-PID controllers have emerged as promising solutions. This chapter reviews extraction methods, associated challenges, and emphasizes the importance of advanced control systems in improving efficiency, product quality, and industrial scalability.

1.1 Background

Essential oils (EO) are volatile aromatic compounds derived from plants through extraction processes. They are found in various parts of plants including flowers, bark, wood, leaves, roots, seeds, and resins [1, 2]. Despite their relatively high cost, essential oils have consistently attracted demand due to their wide applications in medicine, food, cosmetics, and wellness industries. This enduring interest reflects the oils' historical and cultural importance, as well as their commercial and therapeutic value.

Z. Mohd Yusoff et al., *Fuzzy Logic, Smart Control, and the Future of Essential Oils*,
Studies in Fuzziness and Soft Computing 442,
https://doi.org/10.1007/978-981-92-2084-7_1

From a medical perspective, essential oils are prized for their therapeutic properties. They have been employed as natural antibiotics, agents for cellular regeneration, and treatments for skin ailments. Furthermore, certain oils have shown promise as complementary therapies in cancer treatment, as natural sedatives, anti-anxiety agents, and digestive aids. For example, patchouli oil has been extensively used in traditional Chinese medicine, where it demonstrates a variety of pharmacological activities [3]. Aromatherapy applications have also gained popularity, harnessing essential oils to manage stress, improve psychological well-being, and support treatments for infectious diseases.

Beyond medicine, essential oils as in Fig. 1.1 have established roles across industries. They are vital to the perfume and fragrance sectors, pharmaceuticals, and food production [4–6]. For instance, lavender essential oil is widely used in flavoring beverages, confectionery, baked goods, and chewing gum [6]. With Malaysia's vast rainforest resources, the nation holds strong potential for the large-scale commercialization of essential oils, leveraging both biodiversity and increasing global demand.

Fig. 1.1 Common essential oils

1.2 The Challenges

Among the many extraction techniques available, steam distillation remains the most widely used for producing essential oils from botanical materials [7–10]. Compared with methods such as solvent extraction, expression, or supercritical fluid extraction, steam distillation accounts for approximately 93% of global essential oil production [1]. The approach is popular not only because of its historical roots but also due to its relatively low cost and adaptability for industrial-scale operations [11–13].

However, steam distillation has several drawbacks. A typical extraction process requires at least three hours to complete [10, 14], and in some cases considerably longer [14, 15]. Such lengthy operations can result in the loss of volatile and water-soluble compounds [1, 16]. Extended exposure to high temperatures also leads to thermal degradation, altering the aromatic profile, reducing color quality, and producing undesirable off-flavors.

Another limitation is the lack of process selectivity. Traditionally, steam distillation has been operated without careful regulation of steam temperature or heat supply [7, 8]. Yet, temperature is among the most critical parameters influencing both the yield and quality of extracted oils [17, 18]. Excessive heating can cause biomass charring, pyrolysis, and decomposition of essential oil compounds [19]. These issues, while prominent in steam distillation, are also shared by other extraction methods, highlighting the broader challenge of balancing process efficiency with product quality.

1.3 Essential Oils and its Applications

Essential oils sometimes called volatile oil are aromatic substances [8, 20] represent a small friction of plant's composition, which can be found mostly in leaves and flowers [21]. It is multi-component chemicals that will produce a product that is highly aromatic and more concentrated than the original plant. As a rule, the oils composed of a complex mixture of terpenes, oxygenated terpenes, sesquiterpenes and oxygenated sesquiterpenes [1, 2, 22] Essential oils and extracts obtained from various botanical plants have recently gained a great popularity and science interest.

Essential oils have a wide range of uses and promise various significant benefits to human and industry such as in aromatic and fragrances [1, 3], medical activities [16, 23] as insect repellents and pesticides [24, 25], nutraceutical or cosmetics [3, 6, 26], as a natural additive for food flavoring [1–28], anti-tumor activity [24], antibacterial [21, 27, 28], anti-cancer, anti-mutation [21], anti-fungal [27, 28], antimicrobial activity [28], pharmaceutical industries [2, 23, 29], and soil enrichment [24] according to the composition of the oil.

1.4 Extraction Techniques

Essential oil extraction means the process of extracting the aromatic substances from botanical material, which is stored in the pockets of the plants by either breaking or heating the pockets. The essential oil can be isolated through various means from conventional to advanced methods. There are several extraction techniques such as:

- Hydro distillation [1, 2, 9, 19, 26, 30].
- Soxhlet extraction [10, 31].
- Solvent extraction [1, 9, 22, 32, 33].
- Maceration [2].
- Supercritical fluid extraction [2, 9, 16, 21, 23, 25, 29, 33, 34].
- Supercritical carbon dioxide [1, 3, 16, 21, 26, 27, 33].
- Cold pressing [35].
- Superheated water extraction [11, 36, 37].
- Ultrasound-assisted distillation [2, 23, 34].
- Steam distillation [1, 2, 7–9, 16, 19, 22, 23, 29, 32, 34, 38].

Other extraction techniques that have been used in extracting the essential oils such as microwave hydro diffusion and gravity [21, 30, 39], microwave accelerated steam distillation, microwave-assisted distillation, solvent-free microwave extraction, head space co-distillation. These are alternative techniques to the conventional procedures to improve the performance of extraction.

The hydro distillation is the extraction method where the botanical material was placed in a packed bed and completely immersed into the water. The heat was directly applied to it. When the mixture boiled, steam that acts as a transportation medium carries tiny particles of essential oil released. Then, the steam went through the condenser and converts it to produce hydrosol (a mixture of oil and water) [38]. This extraction process provides low cost [27] but thermal degradation of unsaturated compounds, losses of some volatile aromatic compounds, affect the percentage of oil yield composition, hydrolysis and water solubilization of some fragrance constituents may be encountered using this method [1, 2, 23, 27]. Finally, essential oils undergo chemical alteration, and the heat-sensitive compounds can easily be destroyed [40].

Soxhlet extraction is one of the most frequent traditional methods that used to extract organic materials from solid or liquid matrices [10]. Soxhlet apparatus consists of condenser, soxhlet extractor, thimble-holder and distillation task [41]. The raw material is placed in a thimble-holder and filled with condensed fresh solvent from a distillation flask. When the liquid reaches the overflow level, a siphon aspirates the solution of the thimble-holder and unloads it back into the distillation flask, carrying extracted solutes into the bulk liquid. In the solvent flask, the solute is separated from the solvent using distillation. The solute is left in the flask and fresh solvent passes back into the plant solid bed. The operation is repeated until complete extraction is achieved. The main drawbacks of this method are time consuming up to 12 h to complete the extraction, labor intensive [32, 41], Some of the lower molecular weight hydrocarbons were found to be absent in soxhlet extract because they were lost during the extraction or in the solvent reduction steps [10, 37], large quantities of solvent used during evaporation/concentration procedure, non-selective extraction [34, 40] and the possibility of thermal decomposition of the target compounds cannot be ignored as the extraction usually occurs at the boiling point of the solvent for a long-time [40].

Solvent extraction is the classical extraction method to isolate the essential oil from plant material. Countless studies reported the drawbacks using this technique such as the cost not expensive but produce a toxic residue that polluted the foods and fragrances, losses of some volatile compound, thermal degradation, low yield, long processing time and large solvent volumes [27, 34, 42]. The composition of the extracted, mostly similar with steam distillation and hydro distillation [17].

Recently, supercritical fluid extraction has received increasing attention over the traditional method like solvent extraction and hydro distillation [1, 29, 43]. This extraction method has been demonstrated to be an alternative technique for isolation of organic compounds from various plants [29, 44].The botanical material is loaded into an extraction vessel, which is equipped with temperature controllers and pressure valves at both inlet and outlet to keep desired extraction conditions. The extraction vessel is pressurized with the fluid by a pump. The fluid and the dissolved compounds are transported to separators, where the salvation power of the fluid is decreased by decreasing the pressure or increasing the temperature of the fluid. The product is then collected via a valve located in the lower part of the separators [41]. Although supercritical extraction can overcome some of the conventional technique's limitations such as hydrolysis and thermal degradation [44] but this method also provides own disadvantages. For example, very small industrial-scale applications due to economics and onerous operating condition [41], compositions of the oil obtained by supercritical fluid extraction and steam distillation are not qualitatively different [1, 27], this method is too slow for commercial use in comparison with steam distillation at 100 °C [11].

The supercritical carbon dioxide is the most expensive modern technologies using carbon dioxide (CO_2) as the solvent to carry the essential oil away from the raw material at high pressure [40]. The CO_2 is an inert, non-flammable, inexpensive, non-explosive, odorless, non-toxic, clean solvent that leaves no solvent residue in the product; however, CO_2 is lacked polarity for the extraction of polar analytes [16, 29, 40, 41]. This technique has not yet been applied to large-scale plants because of the difficulties encountered with the continuous transport of solid material into, through and out of the high-pressure extractor [33]. Although supercritical CO_2 has advantages on low operating temperatures and solvent-free products, but it is difficult to optimize the operating conditions [45] and owing to contamination of the extracted species with supercritical CO_2 [2].

Superheated water extraction also known as subcritical water extraction is one of the well-established methods for the isolation of valuable compounds from aromatic plants [2]. Some numbers of recent studies apply this method in their studies [2, 10, 36, 46].The superheated water term is used to denote the uses of water as an extractant at temperatures between 100 °C and critical point (374 °C) and at a pressure high enough to maintain the liquid state [37, 46]. Previous scientist reported that this method was a more environmentally friendly and enables the rapid extraction, but it also has a shortcoming such as less efficient, formation of undesirable compounds and frequent use of toxic organic solvents; sample-preparation technique with possible contamination, extraction efficiency problems and higher temperature can cause decomposition of the essential oil.

The use an ultrasound method to the extraction of chemical constituents from plant tissues has shown to diminish the danger of thermal degradation, losses of volatile compounds and aid extraction by significantly reducing extraction times [2, 47]. However, the main disadvantage of this method is the potential formation of the free radicals during sonolysis of the solvent, which can produce degradation of some labile compounds by oxidation [2].

Finally, is steam distillation technique. The steam distillation technique is one of the conventional methods and the most preferred extraction method to obtain essential oils. It is still relevant and preferred in both industrial and research area because of factors such as an operational cost, cleanliness, system cost, productivity and maintenance cost [11, 25]. Many of the newly developed extraction techniques used the steam distillation as a yardstick to compare the quality and quantity of essential oils [6, 11, 16, 19]. The relative chemical composition of volatile fractions obtained by steam distillation and sub critical water extraction (SWE) were relatively similar [36]. Other authors reported that, the loss of some volatile compound, low extraction efficiency, thermal degradation problems can be resolved [36]. Superheated and soxhlet are more effective compared to the steam distillation, but the volume produces is relatively small. Steam distillation also compared with supercritical fluid extraction (SFE) and the result showed that both techniques extract were similar in terms of qualitative [48]. Comparison of all experiments under the chosen conditions shows steam distillation to be the most effective extraction method compared with SFE and superheated water extraction [11]. Compared to the supercritical carbon dioxide (SCO_2), the steam distillation produces more yields but extract similarity of qualitative properties [40]. It proved that the steam distillation extraction process is still reliable alongside with the modern technique and be the primary choice to operate in mass scale for the extraction process in industries.

The steam distillation process as shown in Fig. 1.2 in the extraction of essential oils from botanical material consists of 3 basic main parts, which are the boiler, the condenser, and the separator. The botanical material was placed in the packed bed, sits above of the boiler. The water is heated by immersion heater to produce pressurized steam. The steam passes through the plant material heating and makes the drops of plant oil burst and evaporates. The resulting vapor, a mixture of steam and essential oil vapors, passes out at the top and is conveyed to the condenser. Essential oils do not mix well with water, so it forms a thin film on top of the extracts [9, 36].

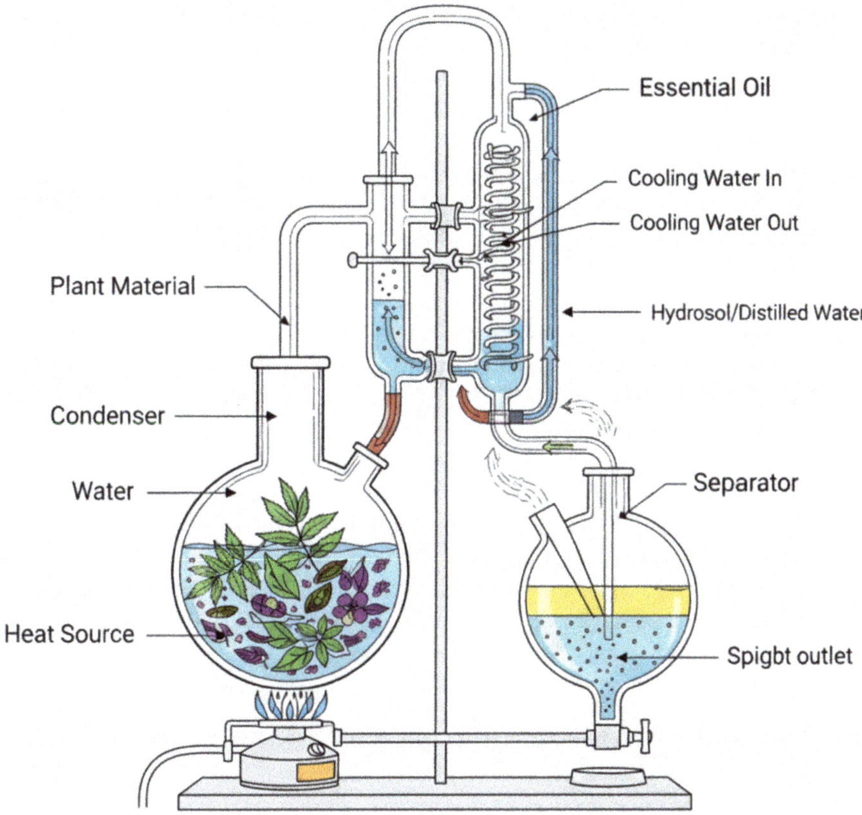

Fig. 1.2 Conventional steam distillation process

1.5 Quality Assessment of Essential Oil

International and national standard organizations such as International Standards Organization (ISO), British Standards (BS), Essential Oil Association of USA (EOA), American Society for Testing and materials (ASTM), Turkish Standards Institute (TSE), etc. has established the quality criteria due to the great variation encountered in the quality of essential oil. Commercial essential oil needs to fulfill this standard before going to the market [49].

The assessment of essential oil quality can be classified into four methods, which are sensory evaluations, physical tests, chemical tests, and instrumental techniques. Sensory evaluations can be carried out by expert nose and after systematic olfactive training. However, sensory evaluation method is very subjective and may vary from one person to another. Thus, the assessment using this method should be verified by a panel of experts and documented by experimental proof. The physical test conducted such as moisture content, specific gravity, optical rotation, refractive index, residue on evaporation, freezing or congealing point, solubility in dilute alcohol. Chemical tests include determination of acid value, carbonyl value, ester value, and ester value after acetylation in some essential oil (anethole in anise and star anise oils; carvone in caraway, dill and spearmint oils; cineole in cardamom, rosemary and sage oils; linalool in coriander and basil oils).

Instrumental techniques include chromatographic and spectroscopic techniques or combination of both techniques (often described as hyphenated techniques). This book only will concentrate on the Gas Chromatography (GC) and Gas Chromatography–Mass Spectrometry (GC-MS) analysis. The GC analysis is the most important technique to separate the constituents of essential oil. It's based on separation of volatile components from a liquid or gas mixture which is volatilized in a column packed. The nitrogen or helium gas was used to carry the volatile compounds through the column. The column was placed in the oven, and the temperature programmed was set. The components within a mixture are separated due to the differences among the partition coefficients of the component between the gas phase and the stationary liquid phase. Compounds which have less affinity to the stationary phase are eluted first from the column and appear as a series of peak. Each peak in the chromatogram represents at least one constituent, and each constituent can be identified by measuring its positions relative to a standard peak of a known substance. The GC safely and routinely used by hyphenated techniques such as GC/MS, GC/FT-IR, NMR etc. [49].

1.6 Extraction of Volatile Oil from Kaffir Lime Peels

Kaffir lime from the Rutaceae family contains two essential oils, makrut leaf oil and makrut (fruit peel) oil [50]. Kaffir lime is a well-known herb used throughout the industry and get high demand in commercialization such as aromatherapy, potential use in medicine, cosmetic industry in making anti-dandruff shampoo and in the food industry as an ingredient in processing chili paste [50–52]. The compounds in the kaffir lime peel such as α-pinene, camphene, β-pinene, sabinene, myrcene, limonene, ɣ-terpinene, terpinolene, trans-sabinene hydrate, copaene, linalool, citronella, terpinen-4-ol, citronellal, geraniol, and δ-cadinene with the refractive index (RI) is 1.4729 [53].

1.7 Extraction with Temperature Controls Efforts

The selection of manipulated variable and process variable of a particular control system is very important because these parameters can significantly influence production yield and chemical composition of essential oil. There are some parameters under considerations such as temperature, pressure, flow rate, extraction time, particle size of extract material, and raw state material (dry or natural) [11, 17, 20, 24, 25, 44]. The parameters under study showed different influences on the volatile oil extracted

[44]. Some previous studies on essential oils stressed that temperature, extraction time, and extract material condition have the most significant effect of the production yield [9, 17, 18, 40, 41, 54, 55]. But pressure and flow rate give less influence on output yield compared to the temperature, extraction time and size of material [56, 57]. Temperature factor is not limited to steam distillation only but on the other extraction techniques. This variable is an important factor regarding oil extraction because some of the material constituting these types of oil is thermolabile and/or bound to suffer hydrolysis. Controlling the temperature for the extraction is important since a slight variation of a few degrees may result in big losses of oil, as well as in oils with very different chemical characteristics, making them either efficient or not, according to the purpose they will serve [38].

1.8 Toward Intelligent Process Control

The extraction of essential oils through steam distillation is a thermally driven process, highly sensitive to variations in operating parameters. Among these, temperature regulation is widely recognized as the most critical factor influencing oil quality and yield. Poorly controlled heating leads to two major challenges:

(i) Overheating: causing pyrolysis, degradation of thermally unstable compounds, and the production of unwanted by-products.
(ii) Underheating: resulting in incomplete vaporization of volatile constituents and, consequently, lower extraction yields.

Traditionally, essential oil distillation has been performed manually, with operators controlling the heating source based on experience and intuition. While this artisanal approach has been practiced for centuries, it is unsuitable for modern industrial production, which demands repeatability, efficiency, and consistent quality standards. This gap highlights the need for intelligent process control systems [38].

In many small- and medium-scale distillation setups, on–off control or simple thermostats are employed to regulate the heating element. Although inexpensive and easy to implement, these systems suffer from oscillations, slow response times, and lack of precision. The result is frequent overshooting or undershooting of the desired temperature set point.

A more advanced alternative is the Proportional-Integral-Derivative (PID) controller, one of the most widely used control methods in industrial processes [58, 59]. PID controllers regulate process variables by continuously adjusting the control signal based on proportional (P), integral (I), and derivative (D) terms. Properly tuned PID controllers can achieve stable and accurate control. However, they have inherent limitations:

- Nonlinearity Handling: PID assumes a relatively linear system, while essential oil distillation involves nonlinear dynamics due to phase change (liquid–vapor transition), biomass variability, and fluctuating heat transfer rates.

- Parameter Sensitivity: Once tuned, a PID controller may not adapt well to sudden changes in load, feedstock properties, or environmental conditions.
- Lack of Adaptability: Conventional PID does not self-adjust, requiring manual retuning when system dynamics vary.

Therefore, while PID control is an improvement over manual or on–off regulation, it is still insufficient for robust, real-time essential oil extraction where plant material heterogeneity and process disturbances are inevitable.

1.9 Emergence of Intelligent Control Systems

To overcome the drawbacks of traditional controllers, intelligent control methods have been increasingly applied in process industries [60–62]. These include:

- Fuzzy Logic Control (FLC): Inspired by human reasoning, FLC is capable of handling uncertainties, imprecise inputs, and nonlinearities. Unlike PID, fuzzy controllers do not require an exact mathematical model of the system. Instead, they rely on linguistic rules ("if–then" statements) that can mimic the decision-making of experienced operators. For instance, a fuzzy controller can reason that "if temperature is slightly above the set point and rising quickly, then reduce heating moderately."
- Neural Networks and Adaptive Control: Neural network-based controllers can learn process behavior through training data, while adaptive control strategies adjust parameters dynamically in response to changing system conditions. However, these approaches often require high computational resources and are less practical in cost-sensitive setups [63, 64].
- Hybrid Controllers (Fuzzy-PID): A promising approach is to combine the reliability of PID with the adaptability of fuzzy logic. A self-tuning fuzzy PID controller can automatically adjust PID parameters based on fuzzy inference rules. This integration enhances robustness, minimizes steady-state error, and maintains optimal performance under varying process conditions [61, 65, 66].

Beyond their technical advantages, intelligent control systems also play a transformative role in the essential oil industry. Many producers still rely on manual distillation techniques that, while preserving cultural heritage, often lead to inconsistencies in product quality and inefficiencies in resource use. Intelligent control technologies offer a practical bridge between traditional practices and modern industrial requirements.

For small-scale or cottage-industry producers, the adoption of cost-effective intelligent controllers can bring significant advantages. It enables better reproducibility of the extraction process and ensures compliance with international quality standards, which is crucial for market acceptance. At the same time, intelligent control reduces wastage of raw biomass and energy, making the production process more resource efficient. Perhaps most importantly, it empowers smaller producers to compete with larger manufacturers by emphasizing product quality rather than sheer production scale. In this way, intelligent process control is not merely a technological upgrade but a strategic pathway toward sustainable commercialization, industrial scalability, and global competitiveness in the essential oil sector.

1.10 Wrapping up

In conclusion, essential oils remain highly valuable natural products with diverse applications across medicine, food, cosmetics, and wellness industries. While steam distillation continues to dominate extraction practices due to its accessibility and historical roots, the method faces significant challenges in efficiency, selectivity, and product consistency. Conventional control approaches, ranging from manual operation to PID regulation, are often insufficient to address the nonlinear and dynamic nature of the extraction process. The emergence of intelligent control systems offers a promising solution, bridging traditional artisanal methods with modern industrial standards. By enhancing process stability, improving quality reproducibility, and reducing resource wastage, intelligent controllers pave the way for sustainable growth and competitiveness of the essential oil industry. Building upon this foundation, the following chapter will delve deeper into the principles, design, and implementation of intelligent control strategies tailored for essential oil extraction.

References

1. Ebrahimzadeh, H., Yamini, Y., Sefidkon, F., Chaloosi, M., Pourmortazavi, S.M.: Chemical composition of the essential oil and supercritical CO_2 extracts of *Zataria multiflora* Boiss. Food Chem. **83**(3), 385–391 (2003). https://doi.org/10.1016/S0308-8146(03)00096-7
2. Roldán-Gutiérrez, J.M., Ruiz-Jiménez, J., Luque de Castro, M.D.: Ultrasound-assisted dynamic extraction of valuable compounds from aromatic plants and flowers as compared with steam distillation and superheated liquid extraction. Talanta. **75**(5), 1369–1375 (2008). https://doi.org/10.1016/j.talanta.2008.01.057
3. Donelian, A., Carlson, L.H.C., Lopes, T.J., Machado, R.A.F.: Comparison of extraction of patchouli (*Pogostemon cablin*) essential oil with supercritical CO_2 and by steam distillation. J. Supercrit. Fluids 48(1), 15–20 (2009). https://doi.org/10.1016/j.supflu.2008.09.020
4. Yusoff, Z.M., Ismail, N.: Datasets of chemical compounds in three different species of Aquilaria using GC-MS coupled with GC-FID analysis. Data Brief. **53**, 110209 (2024). https://doi.org/10.1016/j.dib.2024.110209
5. Mahabob, N.Z., Yusoff, Z.M., Ismail, N., Taib, M.N.: Improving the quality of agarwood oil by using the intelligent of multilayer perceptron (MLP) network training algorithms. In: AIP Conference Proceedings (2023). https://doi.org/10.1063/5.0115568
6. Chemat, F., Lucchesi, M.E., Smadja, J., Favretto, L., Colnaghi, G., Visinoni, F.: Microwave accelerated steam distillation of essential oil from lavender: a rapid, clean and environmentally friendly approach. Anal. Chim. Acta. **555**(1), 157–160 (2006). https://doi.org/10.1016/j.aca.2005.08.071
7. Muhammad, Z., Yusoff, Z.M., Rahiman, M.H.F., Taib, M.N.: Steam temperature control for steam distillation pot using model predictive control. In: Proceedings of the 2012 IEEE 8th

International Colloquium on Signal Processing and Its Applications (CSPA 2012) (2012). https://doi.org/10.1109/CSPA.2012.6194772

8. Yusoff, Z.M., Muhammad, Z., Rahiman, M.H.F., Taib, M.N.: ARX modeling for down-flowing steam distillation system. In: Proceedings of the 2012 IEEE 8th International Colloquium on Signal Processing and Its Applications (CSPA 2012) (2012). https://doi.org/10.1109/CSPA.2012.6194774
9. Masango, P.: Cleaner production of essential oils by steam distillation. J. Clean. Prod. **13**(8), 833–839 (2005). https://doi.org/10.1016/j.jclepro.2004.02.039
10. Ozel, M.Z., Kaymaz, H.: Superheated water extraction, steam distillation and Soxhlet extraction of essential oils of *Origanum onites*. Anal. Bioanal. Chem. **379**(7–8), 1127–1133 (2004). https://doi.org/10.1007/s00216-004-2671-5
11. Ammann, A., Hinz, D.C., Addleman, R.S., Wai, C.M., Wenclawiak, B.W.: Superheated water extraction, steam distillation and SFE of peppermint oil. Fresenius' J. Anal. Chem. **364**(7), 650–653 (1999). https://doi.org/10.1007/s002160051406
12. Huzir, S.M.H.M., Al-Hadi, A.H.H., Yusoff, Z.M., Ismail, N., Taib, M.N.: Accurate Agarwood oil quality determination: a breakthrough with artificial neural networks and the Levenberg-Marquardt Algorithm. IEEE Access. **12**, 50389–50403 (2024)
13. Yusoff, Z.M., Muhammad, Z., Rahiman, M.H.F., Taib, M.N.: PRBS perturbation and modeling of hydro-diffusion steam distillation essential oil extraction system under feed-in disturbance. In: Proceedings of the 2012 IEEE 8th International Colloquium on Signal Processing and Its Applications (CSPA 2012) (2012). https://doi.org/10.1109/CSPA.2012.6194773
14. Deng, C., Xu, X., Yao, N., Li, N., Zhang, X.: Rapid determination of essential oil compounds in *Artemisia selengensis* Turcz by gas chromatography-mass spectrometry with microwave distillation and simultaneous solid-phase microextraction. Anal. Chim. Acta. **556**(2), 289–294 (2006). https://doi.org/10.1016/j.aca.2005.09.038
15. Song, G., Deng, C., Wu, D., Hu, Y.: Comparison of headspace solid-phase microextraction with solvent extraction for the analysis of the volatile constituents of leaf twigs of Chinese arborvitae. Chromatographia. **58**(11–12), 773–778 (2003). https://doi.org/10.1365/s10337-003-0108-5
16. Khajeh, M., Yamini, Y., Shariati, S.: Comparison of essential oils compositions of *Nepeta persica* obtained by supercritical carbon dioxide extraction and steam distillation methods. Food Bioprod. Process. **88**(2–3), 227–232 (2010). https://doi.org/10.1016/j.fbp.2008.11.003
17. Li, X.M., Tian, S.L., Pang, Z.C., Ji, J.Y., Feng, Z.S., Zhang, Y.M.: Extraction of *Cuminum cyminum* essential oil by combination technology of organic solvent with low boiling point and steam distillation. Food Chem. **115**(3), 1114–1119 (2009). https://doi.org/10.1016/j.foodchem.2008.12.091
18. Silou, T., Mapola, G., Makany, R.A., Loumouamou, A.N., Malanda, M., Chalchat, J.C.: Model of steam and water extraction of essential oil of *Eucalyptus citriodora* using a complete 2^n factorial plan. Int. J. Food Eng. **5**(4) (2009). https://doi.org/10.2202/1556-3758.1442
19. Rouatbi, M., Duquenoy, A., Giampaoli, P.: Extraction of the essential oil of thyme and black pepper by superheated steam. J. Food Eng. **78**(2), 708–714 (2007). https://doi.org/10.1016/j.jfoodeng.2005.11.010
20. Arce, A., Pobudkowska, O., Rodríguez, O., Soto, A.: Citrus essential oil terpenless by extraction using 1-ethyl-3-methylimidazolium ethylsulfate ionic liquid: effect of the temperature. Chem. Eng. J. **133**(1–3), 203–207 (2007). https://doi.org/10.1016/j.cej.2007.01.035
21. Bousbia, N., Abert Vian, M., Ferhat, M.A., Petitcolas, E., Meklati, B.Y., Chemat, F.: Comparison of two isolation methods for essential oil from rosemary leaves: hydrodistillation and microwave hydrodiffusion and gravity. Food Chem. **114**(1), 355–362 (2009). https://doi.org/10.1016/j.foodchem.2008.09.106
22. Marongiu, B., Piras, A., Porcedda, S.: Comparative analysis of the oil and supercritical CO_2 extract of *Elettaria cardamomum* (L.) Maton. J. Agric. Food Chem. **52**(20), 6232–6236 (2004). https://doi.org/10.1021/jf034819i
23. Lucchesi, M.E., Chemat, F., Smadja, J.: Solvent-free microwave extraction of essential oil from aromatic herbs: comparison with conventional hydro-distillation. J. Chromatogr. A. **1043**(2), 323–327 (2004). https://doi.org/10.1016/j.chroma.2004.05.083

24. Sayyar, S., Abidin, Z.Z., Yunus, R., Muhammad, A.: Extraction of oil from Jatropha seeds-optimization and kinetics. Am. J. Appl. Sci. **6**(7), 1390–1395 (2009). https://doi.org/10.3844/ajassp.2009.1390.1395
25. Cassel, E., Vargas, R.M.F.: Experiments and modeling of the *Cymbopogon winterianus* essential oil extraction by steam distillation (2006)
26. Marongiu, B., Piras, A., Porcedda, S., Scorciapino, A.: Chemical composition of the essential oil and supercritical CO_2 extract of *Commiphora myrrha* (Nees). Engl. and of *Acorus calamus* L. J. Agric. Food Chem. **53**(20), 7939–7943 (2005). https://doi.org/10.1021/jf051100x
27. Guan, W., Li, S., Yan, R., Tang, S., Quan, C.: Comparison of essential oils of clove buds extracted with supercritical carbon dioxide and other three traditional extraction methods. Food Chem. **101**(4), 1575–1581 (2007). https://doi.org/10.1016/j.foodchem.2006.04.009
28. Chanthaphon, S., Chanthachum, S., Hongpattarakere, T.: Antimicrobial activities of essential oils and crude extracts from tropical *Citrus* spp. against food-related microorganisms. Songklanakarin J. Sci. Technol. **30**(Suppl. 1), 125–131 (2008)
29. Giray, E.S., Kirici, S., Kaya, D.A., Türk, M., Sönmez, Ö., Inan, M.: Comparing the effect of sub-critical water extraction with conventional extraction methods on the chemical composition of *Lavandula stoechas*. Talanta. **74**(4), 930–935 (2008). https://doi.org/10.1016/j.talanta.2007.07.040
30. Vian, M.A., Fernandez, X., Visinoni, F., Chemat, F.: Microwave hydrodiffusion and gravity, a new technique for extraction of essential oils. J. Chromatogr. A. **1190**(1–2), 14–17 (2008). https://doi.org/10.1016/j.chroma.2008.02.08
31. Hashim, Y.Z.H.-Y., Jamil, M.A.M., Jamal, P., Zainurin, N.A.A., Azziz, S.S.S.A.: Hydrodistillation and Soxhlet extraction of Agarwood leaf extract from *Aquilaria malaccensis*. Malays. J. Fundamental Appl. Sci. **15**(6), 842–846 (2019). https://doi.org/10.11113/mjfas.v15n6.1589
32. Peng, F., Sheng, L., Liu, B., Tong, H., Liu, S.: Comparison of different extraction methods: steam distillation, simultaneous distillation and extraction and headspace co-distillation, used for the analysis of the volatile components in aged flue-cured tobacco leaves. J. Chromatogr. A. **1040**(1) (2004). https://doi.org/10.1016/j.chroma.2004.03.057
33. Kiriamiti, H.K., Rascol, E., Marty, A., Condoret, J.S.: Extraction rates of oil from high oleic sunflower seeds with supercritical carbon dioxide. Chem. Eng. Process. **41**(8) (2002). https://doi.org/10.1016/S0255-2701(01)00191-X
34. Glausch, A., Blanch, G.P., Schurig, V.: Enantioselective analysis of chiral polychlorinated biphenyls in sediment samples by multidimensional gas chromatography-electron-capture detection after steam distillation-solvent extraction and sulfur removal. J. Chromatogr. A. **723**(2) (1996). https://doi.org/10.1016/0021-9673(95)00846-2
35. Rahiman, M.H.F., Taib, M.N., Salleh, Y.M.: Performance of multi-step-ahead-prediction ARX for steam temperature in a self-refilling distillation essential oil extraction system. In: ICCAS 2007—International Conference on Control, Automation and Systems (2007). https://doi.org/10.1109/ICCAS.2007.4406608
36. Özel, M.Z., Göğüş, F., Hamilton, J.F., Lewis, C.: Analysis of volatile components from *Ziziphora taurica* subsp. *taurica* by steam distillation, superheated-water extraction, and direct thermal desorption with GC×GC-TOFMS. Anal. Bioanal. Chem. **382**(1), 115–119 (2005). https://doi.org/10.1007/s00216-005-3156-x
37. Özel, M.Z., Clifford, A.A.: Superheated water extraction of fragrance compounds from *Rosa canina*. Flavour Fragr. J. **19**(4), 354–359 (2004). https://doi.org/10.1002/ffj.1317
38. Yusoff, Z.M.: Self-tuning Fuzzy PID Control of Hydro-diffusion Essential Oil Extraction System. Universiti Teknologi MARA, Malaysia (2014)
39. Cendres, A., Chemat, F., Maingonnat, J.F., Renard, C.M.G.C.: An innovative process for extraction of fruit juice using microwave heating. LWT. **44**(4), 1035–1041 (2011). https://doi.org/10.1016/j.lwt.2010.11.028
40. Pourmortazavi, S.M., Hajimirsadeghi, S.S.: Supercritical fluid extraction in plant essential and volatile oil analysis. J. Chromatograph. A. **1163**(1–2), 2–24 (2007). https://doi.org/10.1016/j.chroma.2007.06.021

41. Wang, L., Weller, C.L.: Recent advances in extraction of nutraceuticals from plants. Trends Food Sci. Technol. **17**(6), 300–312 (2006)
42. Kristiawan, M., Sobolik, V., Allaf, K.: Isolation of Indonesian cananga oil using multi-cycle pressure drop process. J. Chromatogr. A. **1192**(2), 308–318 (2008). https://doi.org/10.1016/j.chroma.2008.03.068
43. Yamini, Y., Khajeh, M., Ghasemi, E., Mirza, M., Javidnia, K.: Comparison of essential oil compositions of *Salvia mirzayanii* obtained by supercritical carbon dioxide extraction and hydrodistillation methods. Food Chem. **108**(1), 341–346 (2008). https://doi.org/10.1016/j.foodchem.2007.10.036
44. Grosso, C., Ferraro, V., Figueiredo, A.C., Barroso, J.G., Coelho, J.A., Palavra, A.M.: Supercritical carbon dioxide extraction of volatile oil from Italian coriander seeds. Food Chem. **111**(1), 197–203 (2008). https://doi.org/10.1016/j.foodchem.2008.03.031
45. Raeissi, S., Diaz, S., Espinosa, S., Peters, C.J., Brignole, E.A.: Ethane as an alternative solvent for supercritical extraction of orange peel oils. J. Supercrit. Fluids. **45**(3), 311–317 (2008). https://doi.org/10.1016/j.supflu.2008.01.008
46. Gogus, F., Ozel, M.Z., Lewis, A.C.: Superheated water extraction of essential oils of *Origanum micranthum*. J. Chromatogr. Sci. **43**(2), 87–91 (2005). https://doi.org/10.1093/chromsci/43.2.87
47. Kimbaris, A.C., Siatis, N.G., Daferera, D.J., Tarantilis, P.A., Pappas, C.S., Polissiou, M.G.: Comparison of distillation and ultrasound-assisted extraction methods for the isolation of sensitive aroma compounds from garlic (*Allium sativum*). Ultrason. Sonochem. **13**(1), 54–60 (2006). https://doi.org/10.1016/j.ultsonch.2004.12.003
48. Scalia, S., Giuffreda, L., Pallado, P.: Analytical and preparative supercritical fluid extraction of Chamomile flowers and its comparison with conventional methods. J. Pharm. Biomed. Anal. **21**(3), 549–558 (1999). https://doi.org/10.1016/S0731-7085(99)00152-1
49. De Silva, K.T., Bhat, K.K.S., Lawrence, B.M., Baser, K.H.C., Brud, W.S., Verlet, N.: A manual on the essential oil industry. In: Proceedings of the 3rd UNIDO Workshop on Essential Oil and Aroma Chemical Industries (1995)
50. Srisukh, V., Palliyaguru, L., Sotanaphun, U., Sookkongwaree, K., Wannasiri, S., Ruangrungsi, N.: Antibacterial activity of essential oils from *Citrus hystrix* (makrut lime) against respiratory tract pathogens. ScienceAsia. **38**(2), 212–217 (2012). https://doi.org/10.2306/scienceasia1513-1874.2012.38.212
51. Haiyee, Z.A., Winitkitcharoen, C.: Extraction of volatile oil from kaffir lime leaves (*Citrus hystrix*) using pressurised liquid extraction. Int. J. Food Nutr. Public Health. **5**(1-2–3), 10.47556/j.ijfnph.5.1-2-3.2012.12 (2012)
52. Foo-trakul, P., Watchiradatsatiean, C.: Development of Anti-dandruff shampoo from Kaffir Lime which is the by-product of food industry. Kasetsart J. (Nat. Sci.) **39**, 725–729 (2005)
53. Lawrence, B.M., Hogg, J.W., Terhune, S.J., Podimuang, V.: Constituents of the leaf and peel oils of *Citrus hystrix*, D.C. Phytochemistry. **10**(6), 1307–1316 (1971). https://doi.org/10.1016/s0031-9422(00)84357-5
54. Kéïta, S.M., Vincent, C., Schmit, P., Ramaswamy, S., Bélanger, A.: Effect of various essential oils on *Callosobruchus maculatus* (F.) (Coleoptera: Bruchidae). J. Stored Prod. Res. **36**(4), 355–364 (2000). https://doi.org/10.1016/S0022-474X(99)00055-7
55. Eikani, M.H., Golmohammad, F., Rowshanzamir, S.: Subcritical water extraction of essential oils from coriander seeds (*Coriandrum sativum* L.). J. Food Eng. **80**(2), 735–740 (2007). https://doi.org/10.1016/j.jfoodeng.2006.05.015
56. Rezzoug, S.A.: Optimization of steam extraction of oil from maritime pine needles. J. Wood Chem. Technol. **29**(2), 166–186 (2009). https://doi.org/10.1080/02773810902879025
57. Louli, V., Folas, G., Voutsas, E., Magoulas, K.: Extraction of parsley seed oil by supercritical CO_2. J. Supercrit. Fluids. **30**(2), 163–174 (2004). https://doi.org/10.1016/j.supflu.2003.07.003
58. Soyguder, S., Karakose, M., Alli, H.: Design and simulation of self-tuning PID-type fuzzy adaptive control for an expert HVAC system. Expert Syst. Appl. **36**(3 Part 1), 4586–4600 (2009). https://doi.org/10.1016/j.eswa.2008.05.031

59. Tang, K.S., Man, K.F., Chen, G., Kwong, S.: An optimal fuzzy PID controller. IEEE Trans. Ind. Electron. **48**(4), 751–759 (2001). https://doi.org/10.1109/41.937407
60. Majeed, W.S., Nasser, A.I., Hameed, K.R.: Improve the performance of automatic voltage regulator for power system using self-tuning fuzzy-PID controller. Indones. J. Electr. Eng. Comput. Sci. **29**(3), 1247–1257 (2023). https://doi.org/10.11591/ijeecs.v29.i3.pp1247-1257
61. Chen, W.C., Chang, N.B., Chen, J.C.: Rough set-based hybrid fuzzy-neural controller design for industrial wastewater treatment. Water Res. **37**(1), 95–107 (2003). https://doi.org/10.1016/S0043-1354(02)00255-5
62. Lee, C.C.: Fuzzy logic in control systems: fuzzy logic controller—Part I. IEEE Trans. Syst. Man Cybern. **20**(2), 404–418 (1990). https://doi.org/10.1109/21.52551
63. Kriegeskorte, N., Golan, T.: Neural network models and deep learning. Curr. Biol. **29**(7), R231–R236 (2019). https://doi.org/10.1016/j.cub.2019.02.034
64. Bortnik, J., et al.: Artificial neural networks for determining magnetospheric conditions. In: Camporeale, I., Wing, S., Johnson, J. (eds.) Machine Learning Techniques for Space Weather, pp. 279–300. Elsevier, Amsterdam (2018). https://doi.org/10.1016/B978-0-12-811788-0.00011-1
65. Erenoglu, I., Eksin, E., Yesil, E., Guzelkaya, M.: An intelligent hybrid fuzzy PID controller. In: 20th European Conference on Modelling and Simulation: Modelling Methodologies and Simulation Key Technologies in Academia and Industry (ECMS 2006) (2006). https://doi.org/10.7148/2006-0062
66. Kasuan, N., Yusuf, Z., Taib, M.N., Rahiman, M.H.F., Tajuddin, N., Aziz, M.A.A.: Robust steam temperature regulation for distillation of essential oil extraction process using hybrid fuzzy-PD plus PID controller. World Acad. Sci. Eng. Technol. **47**, 102–107 (2010)

Chapter 2
System Identification and Controller Design: Unlocking Efficiency in Extraction Systems

Abstract This chapter presents an overview of system identification and controller design approaches for temperature regulation in essential oil extraction processes. Accurate modeling of the extraction system is essential for understanding its dynamic behavior and for developing effective control strategies. The chapter introduces different modeling approaches, including white-box, black-box, and gray-box models, and discusses both linear and nonlinear system identification techniques such as ARX and NARX models. Key aspects of experimental design, data collection, and model validation are also highlighted to ensure reliable model development. In addition, the chapter reviews various control strategies used in thermal processes, ranging from conventional methods such as on–off control and proportional–integral–derivative (PID) controllers to more advanced intelligent control approaches. The limitations of traditional controllers in handling nonlinear and time-varying systems are discussed, followed by an introduction to intelligent control techniques such as fuzzy logic control and hybrid fuzzy–PID controllers. These methods offer improved adaptability and performance in complex extraction systems.

2.1 Modeling for Heating Process

Heating process is a common thermal process in many industrial applications such as food, manufacturing and agricultural [1]. Heating applied in food products with intended to improve quality and to extend shelf life of the products [1]. Heating is used to increase or decrease the temperature in manufacturing [2]. In essential oil extraction system, numerous works have been reported that temperature of heating process contribute a major effect on producing the essential oil [3–6]. Heat is the main source for all essential oil extraction systems and strongly affects the chemical composition and percentage yield. Heating consumes the largest part to the energy required [2, 5, 7]. Consequently, it is very important to keep control of such energy demanding process without spending too much energy [7, 8]. So, in most existing works, modeling and control algorithm was applied.

Z. Mohd Yusoff et al., *Fuzzy Logic, Smart Control, and the Future of Essential Oils*, Studies in Fuzziness and Soft Computing 442,
https://doi.org/10.1007/978-981-92-2084-7_2

Generally, modeling heating process is necessary to fulfill some requirements in many processes such as to explain and understood the observed phenomena, to run production safely and efficiently, predicting energy consumption and analyzing control strategies [9]. Successful in the implementation advanced controller design usually based on a good dynamic model to the process under consideration [10, 11]. There are many modeling approaches used in heating process such as first principal model and empirical model [11–14]. Between first principal model and empirical model, lay the semi-physical approach named as gray-box models [8].

The development of the first principal model involves an analytical approach and applies basic laws of physics to explain the dynamic behavior during the process [7, 15]. Model constructed in this way also called white box models or simulation models [7]. For example, in the control strategy analysis of a batch distillation column apply the first principles model to the process to represent the dynamic nature [16]. However, this type of model may not be possible because due to the complexity for practical implementation, costly and the process may not well be understood if the model requires many equations with a significant number of process variables such as physical and chemical properties [11, 17, 18]. A good first principal models is still insufficient to explain the time delay as often seen in the empirical model [19].

The empirical approach is an alternative approach where some experiments are carried out from the system to obtain input–output data [11, 18]. This model does not reflect any physical structure to the system and known as system identification approach or black box model [7]. In the case where the processes are too complex to represent by first principal model because of time consuming, the empirical model comes as an attractive solution [17, 20, 21]. For example, this model shows a very reliable alternative to obtain a mathematical model that may adequately describe the steam pressure variation process in the boiler [21, 22]. The advantage of the black box model is they do not need the previous knowledge about the system and the possibility of obtaining a vague model with a relatively small group of measurements [23, 24]. This technique is easy to construct and adequate for controller tuning purposes [19].

The gray box model is a combination of white box and black box models, which means that a-prior knowledge of the process is used, and unknown parts are estimated from measured data [7, 12, 25, 26]. This approach is suitable when the process is nonlinear because gives the possibility to incorporate a nonlinear description of the model [7]. Several successful applications have been made by using gray box modeling based on mathematical relations such as to estimate physical process parameters of heating process and apply for control purposes [7, 25].

2.2 System Identification of Distillation Column and Related Chemical Process

System identification is a quite mature area that has an interesting and productive element. System identification can be defined as an experimental approach of developing or improving a mathematical model of a dynamic system based upon the relationship between input–output data to describe the underlying mechanism of the systems [15, 21, 27]. The dynamic model is required to design and implementing a high-performance control system. A model can be used as an alternative for testing faults and limitations of expensive and unavailable actual systems [28]. System identification is a well-established technique and especially useful to model the system that cannot represent by first principal model [27, 29]. The implementation of this model increased dramatically and across multi-disciplinary areas. This model has a rapid progress and recognized as an important tool to represent many chemical processes such as a boiler system, tubular flow reactor, food-processing, heat exchanger, tropical greenhouse and electric resistance furnace. In the distillation process, many published works reported that system identification was widespread applied to develop a sophisticated and efficient algorithm [1, 2, 10–12, 21, 30, 31].

Many modeling methods exist for system identification, generally divided as linear and nonlinear. Some common linear black box model such as Finite Impulse Response (FIR), Auto-Regressive with Exogenous Input (ARX) model, Auto Regressive Moving Average with Exogenous input (ARMAX) model, Output-Error (OE) model and Box–Jenkins (BJ) model [15, 20, 32, 33]. Nonlinear Auto-Regressive with Exogenous input (NARX), Artificial Neural Network (ANN), Neural network Auto Regressive with Exogenous input (NNARX), Nonlinear Autoregressive Moving Average (NARMA), and Nonlinear Auto Regressive Moving Average model with exogenous inputs (NARMAX) models are some examples of the nonlinear model [15, 17]. The ARX model is widely used to describe the behavior during the process and brings a lot of advantages due to the simplest structure model, easy to find analytical solutions and provide excellent performance in their implementation [9, 34, 35].

Distillation column is one of the most often unit operations used and the most energy-intensive operations in industrial plants [16, 19, 36]. The types of distillation column are multi-tray column, packed column, and binary column [17, 36]. The tray column is vapor and liquid are contacted on discrete stages, and packed column is contact occurs continuously through the column [17]. Meanwhile, binary column is used to separate the binary mixtures such as water-oil components in essential oil extraction. Identification of the distillation column is not an easy task and seems to be tricky due to large time delay and large time constant [37]. Another difficulty is process nonlinearity and slow dynamic process [33, 35, 38, 39]. But if operated over a sufficiently small region, a control system based on a linear input–output model can be performed satisfactorily.

Nowadays, a large extensive amount of work dealing with the system identification of distillation column for various applications have been reported [17]. In modeling the distillation column, the main control variable is temperature [32, 34,

40–42]. For example, the input–output data of steam temperature collected from self-refilling distillation column was modeled by using ARX model structure. The model is found sufficiently flexible to represent the system dynamic and sufficient for controller design. The ARMAX model structure was implementing to model the steam temperature response that includes the disturbance in distillation column. The ARMAX model was validated to verify its preciseness, and the result shows that the performance criteria using fitness is 98.5%. This model accepted for controller design [43].

The control strategy is applied to dynamic simulations of two-methanol-water distillation column by using neural networks to model the steady-state of the process. Neural networks have been successfully applied to various problems such as system identification, pattern recognition, process modeling, and fault diagnosis. In the area of distillation control, neural networks have found application in identification and control of packed distillation column where a neural network model was used as the model in controller design. The neural network model-based controllers show robust performance for both set points and disturbances and performed better than conventional feedback proportional–integral (PI) controller [8].

A growing literature within the field of chemical engineering applied artificial neural networks (ANNs) in process modeling and control. ANNs were used to model and control of a packed distillation column because can provide good empirical models of complex nonlinear processes that are useful for many purposes, including process control. From the results shown, ANNs can model the column, and demonstrate that the network model is as good or better than a simplified first principles model when used for model predictive control [11, 44–46]. Other several works have been carried out for modeling the dynamic system of distillation column by using different model structure from system identification [17, 36, 47–49].

2.3 System Identification Procedure

The system identification procedure has a natural logical flow. Its starts with system observation and data gathering from experiment that involves the system to be modeled, followed by model structure selection, then model estimation and lastly model validation. If the model obtained did not pass the validation tests, then the procedure must be revised from the beginning or select model structure or just look at another model estimation [15, 21, 50, 51]. If it is concluded that the model may be invalid, the following aspects must be reviewed as possible causes [21, 26, 50].

- The collected input–output data does not informative enough about system dynamics to provide guidance in selecting good models.
- The chosen structure is not able to describe adequately the model.

The system identification approach is an iterative procedure and illustrated in Fig. 2.1, based on the methodologies reported in [21, 23, 50, 51]:

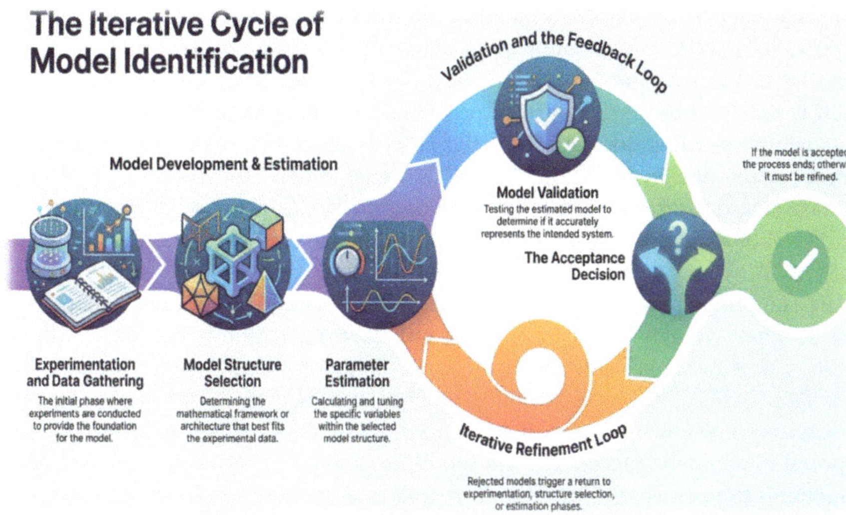

Fig. 2.1 System identification approach

2.3.1 Experiment Design

Experimental design is the basis procedure and become the most important part for the system identification. The input–output data are collected for the full dynamicity of the system during an experiment in real time or simulated using a computer system [26, 50, 51]. The set of data from the system must be maximally informative to ensure the model present is accurate and can describe the dynamic behavior over its entire range of operating [26, 32]. There are two approaches in collecting data, which is open-loop and closed-loop. The open-loop approach is relatively like direct approach closed-loop where the model is identified from the input–output data by ignoring the feedback [52–56].

In system identification, the type of input signals to be used plays a crucial role and has a direct fidelity of the resulting identified model. The excitation signal should have the important characteristic below [29].

- The input signal must be persistently exciting or fluctuating enough to excite all the dynamic modes of interest [29].
- The input signal must have different levels of input amplitudes over the whole region of operation [29].

Various input signals are applied to capture the dynamics within the system such as Pseudo Random Binary Sequence (PRBS), pulse step function, Random Binary Sequence (RBS), white-noise, sine-wave, Gaussian signal, and random multi-level sequence [21, 29, 32, 50, 55, 57–64]. The choices of input signals depend on the intended applications. The PRBS is the most popular input signal and most of the published work on system identification use PRBS as their perturbation signal [32, 52, 65, 59–61, 66]. The advantages of using PRBS signal are its power spectrum is similar to the white noise [15, 27, 61]. Signal varies between only two values, which is practical in most industrial processes, and the signal can be concentrated frequency ranges that correspond to the process dynamics which is important for control system design [18, 52]. The PRBS signal is used for triggering the output response at various frequencies and is an appropriate means to drive the system [43, 58]. The PRBS signals can be generated by using MATLAB System Identification Toolbox.

The sampling interval is one of the important issues during data collection, it must be small enough so that significant process information is not lost. However, sampling at a too small intervals can cause data redundancy while too large will cause data lost [15, 50]. The sampling interval can only be chosen appropriately if there exists some knowledge of the dynamics within the system. Some authors were proposed the guideline to choose the sampling interval is between 1 and 5% of the dominating time constant [18, 50]. A simple rule of thumb is to sample as fast as possible and certainly on the tenth of the dominant time constant [67]. In distillation, since the time constant for steam temperature is very large and based upon a literature regarding sampling interval, 1 s has been chosen as the sampling interval [14, 68].

Collected data from the processes is generally accompanied by undesired noise and other kind of imperfections, which could be necessary to correct before going for further processing [21]. Using raw data is possible but data without pre-processing are sometimes meaningless, the identification results might be an inaccurate and computing algorithm did not converge [15, 50]. There are several types of data pre-processing for extracting the most valuable information from the measured data such as filtering, removing outliers, and magnitude scaling. Filtering is used to remove noise, periodic disturbances and uninteresting process dynamic from the training and testing data set [15] by blocking the rapid changing component to the signal. So that, the filter output will have less noise than does the raw signal. However, filtering may have little or no effect for the slow process. Removing outliers is to remove outliers that affect the estimate model and usually remove before training the data. Meanwhile, magnitude scaling is to scale the variables to the same variance [15].

Numerous published studies, which deal with the identification usually, separated the collected data into several groups. The size of data split is normally according to a ratio of 50:50; represent the estimation data and validation data. Some works dissected the original input-output data into two parts by using interlacing technique where odd sequence data meant for training while even sequence data is meant for validation purposes [20, 32, 43]. In some cases, data were split into three portions, which are training data, validation data and test data. There are also several works where the validation data is used totally from a different set of data obtained from a different experiment [65, 66, 69]. This approach is preferable, especially when data collection is feasible without any cost. Other combinations of data split ratio is also possible, but it is recommended that the training data sufficiently represent the process behavior and cover all operating points [64].

The size and time duration for data collection is varied and depends upon the experiment and processes itself. The experiment should excite at least all states through the process so that the model used is affected by exposure process and cover all the dynamic range of the process [7, 35, 70]. For example, the time to collect the data is chosen to be within the interval 10–60 min [7]; the size of data collection is 3900 samples used for system identification [64], and the system is excited using $N = 5000$ samples of random binary input signals generated by MATLAB *idinput* command [71].

2.3.2 Model Structure Selection

Model structure selection constitutes one of the most important and difficult decisions that must be made since model complexity can affect exactitude as the model describes the real process [21, 50]. At this stage, the appropriate polynomials are determined that will constitute to model framework [21, 72–74]. Nonlinear structure such as NARX, NNARX, neural network, and fuzzy model [12, 27]. The OE model structure gives a good description of low frequency properties to the system; ARX model structure provides a better description of high-frequency characteristic and ARMAX model structure allows modeling of both deterministic and stochastic response to steam pressure variation process [21].

Selection of the model structure is derived using *prior* knowledge and wrongly selecting model structure resulting in a poor model [29]. The nonlinear model is valid for a wide operating range, but ARX model structure is most used in [60].

Once the model structure has been selected, the next step is to select the model order. The choice of the model order is a nontrivial problem for practitioners and researcher [26, 50]. The choice of the model order is a compromise between the unmodeled dynamics and increasing the complexity of the model which can lead to model stabilizability difficulties. The model order can be either chosen by randomly trial-and error (heuristic) or based on numerical approach [50]. Trial-and-error approach is possible and relatively easier, but the chosen model order might not be the optimal order [26, 50]. There are several numerical approaches in determining the model order such as Akaike's final prediction error (FPE), Akaike's information criterion (AIC), Rissanen's minimum description length (MDL), Normalized sum of squared errors (NSSE), and the Lipschitz function [75]. However, in many practical cases, a second order or even a first-order model is adequate [20, 35].

2.3.3 Model Estimation

Model estimation is to estimate the value of unknown parameters for the selected model structure by solving the estimation problem numerically. A good model is a model with good prediction capability or model that produces small prediction error

[50]. Two popular approaches are correlation approach and prediction error approach. The correlation approach is realized by using the least square method and prediction error approach is realized by minimizing a certain loss function [26, 50, 51].

Prediction error method (PEM) is most preferred to estimate a model. Prediction error or also known as residual is the different between generated data and observed data [50]. Prediction error method will minimize the error between the predicted and observed values of the output and performed in the iterative manner. The measurement of the prediction error is often represented by a function known as loss function. Many publications which deal with the identification implement the PEM approach in their works [66, 69, 75, 76].

2.3.4 Model Validation

Model validation played a major role in system identification, it enables the best model to be chosen for a given plant such as the best structure and the best algorithm for the estimation of the parameters, and as the last step quality control station before a model is delivered to the user [50, 68]. This stage is important for the model to demonstrate the same dynamic characteristics as the real process [7]. Two general approaches have been used to validate the estimated model known as a graphical and statistical approach [15]. Graphical approach is simpler and need common sense in interpreting the result while statistical approaches are used to select the satisfactory model and produce result by a statistical method. The validation test is basically based on simple difference between predicted output and measured output. Deficient model behavior makes the model will be rejected while good performance will develop a certain confidence on it [15].

Several tests need to be performed to ensure good model fit and whiteness of the residuals. Among the popular technique employed in validation test is one-step-ahead prediction (1-SAP) [29, 70, 77]. 1-SAP is produced based on the past actual output rather than the past prediction output. The closeness of the 1-SAP output to the observed output is called as the goodness of the model prediction or the model fit [75, 77]. The difference between predicted output and measured output is called prediction error or residual. Residual tests are very useful for statistical validation such as the coefficient of determination (R^2), adjusted-R^2, root mean-square error (RMSE), residual histogram, residual mean and variance and correlation function [50]. The R^2 is usually present in percentage, and high percentage means the model is good [75]. The correlation function consists of auto-correlation function (ACF) and cross-covariance function (CCF). Many published works applied residual test to validate their model [7, 12, 14, 21, 69, 76–78]. The model is only accepted if both fitting and residual tests are passed.

2.4 Controller Design

Control engineers design and install numerous controllers within a process to allow them to meet tough environment emission restrictions, maintain product quality targets, and comply with a certain safety guideline [18]. There are two general approaches to design the controller systems which are traditional approach and model-based approach [18].

- Traditional Approach: The control strategy and control system hardware are selected based on the knowledge about the process, experience and insight. The controller settings are adjusted after the control system is installed at the plant and known as controller tuning. The disadvantages of this approach are the plant design determines the process dynamic as well as the operability of the plant. In extreme situations, the process may be uncontrollable, even though the design appears satisfactory from a steady-state point of view [18].
- Model-Based Approach: Firstly, the dynamic model that describes the behavior during the process was developed that can be helpful in at least three ways: (i) this model can be used as the basis for model-based controller design method; (ii) the dynamic model can be integrated directly in the control law; and (iii) the model can be used as a computer simulation as an alternative control strategy and to determine preliminary results of the controller settings [18].

Several model-based control schemes have been proposed to integrate a process model into a control system to design a high-performance controller [79]. The design and implementation effective control systems for various applications require the dynamic model that adequately describe the real process behavior [21]. In complex processes, a process model is invaluable for control system design and to improve understanding about the process [18]. With system identification techniques, it is possible to obtain models that agree with the physical reality and useful for model-based control systems design [21].

2.4.1 PID Controller

Despite all the advances in control over the past 50 years, the traditional proportional-integral-derivative (PID) controller is still overwhelmingly used in the process industries [80, 81]. This popularity is partly because PID controller is simple architecture and well understood, ease of implementation, and robustness in operation [81–85]. PID controller is usually used as a yardstick to compare with the other controller performance.

PID controller consists of three parameters, which are needed to be tuned properly to obtain optimum performance known as proportional gain, P, integral gain, I and derivative gain, D [86]. The selection of Proportional (P), Integral (I) and Derivative (D) values are very crucial because these three parameters determine whether the process is oscillatory, stable or unstable [87]. The stability of the PID controller is highly depended on its parameter. Many methods have been proposed for tuning this controller parameter such as heuristic methods [88], Ziegler and Nichols method [89–91], Cohen-Coon method [91], Wang's method [86], genetic algorithm [22, 39, 81, 83, 92], gradient descent [89], pole assignment method [83], particle swarm optimization approach [88], and Nelder–Mead method [85].

The Ziegler–Nichols methods are the most popular methods used and as a benchmark in process control to determine the parameters of PID controller. There are two types of Ziegler–Nichols method, which is a continuous cycling method and process reaction curve. Ziegler–Nichols tuning approach requires information of ultimate gain and the ultimate period of oscillation from the step test reaction curve in order to calculate the controller parameters [85, 90]. Ziegler Nichols formula assures good load-disturbances attenuation, but it generally provides a poor phase margin, and therefore, it might produce a large overshoot and settling time for the step response [91, 93, 94]. A common disadvantage of the Ziegler–Nichols methods is the resulting closed-loop system is often more oscillatory than desirable. Generally, it is often hard to determine an optimal or near optimal PID parameter in the Ziegler–Nichols formula in many industrial plants [88, 95].

Unfortunately, it has been quite difficult to tune properly the gains of PID controller because many industrial plants are often burdened with problems such as high order, large time constant, long-time delay, and nonlinearities [38, 85, 88, 96]. Consequently, PID controller might not achieve adequate control performance [96]. PID controllers generally did not work well for nonlinear systems, higher order, particularly complex and systems that have no precise mathematical model [97]. Numerous literatures claimed that the implementation of PID controller is unsuitable for a certain application, for example, PID controller produces a very high overshoot and is unsuitable for batch reactor temperature control [98]; power consumption of PID controller is much higher when compared to the fuzzy controller [99], and PID controller is suffering with settling time issues [100]. The PID controller also cannot always effectively control systems with changing parameters and may need frequent on-line retuning [94]. For these reasons, several advanced control approaches have been presented to improve PID performance such as fuzzy logic controllers, artificial neural networks (ANN), generalized predictive control and adaptive control [52, 81, 88, 89, 96].

2.4.2 Fuzzy Logic Controller

During the past several years, fuzzy logic controller (FLC) which is, firstly, applied by Mamdani in mid-70s has emerged as one of the most active and fruitful areas for research on the application of fuzzy set theory [101, 102]. FLC's are unique controller and normally built based on experience and knowledge of experts to develop the rule of the controller [90, 93, 94, 103]. A fuzzy control provides a formal methodology for representing, manipulating and implementing a human's heuristic knowledge about how to control a system [104]. Effective utilization of fuzzy control has been successfully applied in numerous application such as water quality control, automatic train operation system, fuzzy logic controller hardware systems, elevator control, nuclear reactor control, and fuzzy computers [101]. The FLC is one of the well-succeeded technologies for development of sophisticated process control systems [43].

Experience shows that the FLC yields results superior to those obtained by a conventional controller such as work well for nonlinear systems and uncertainty, higher-order systems, time-delayed linear systems, particularly complex and vague system that has no precise mathematical models [93, 97, 101, 105]. FLC's are more robust and has been proven as an alternative approach to conventional controllers based on flexibility and fast reactions, simplest design, and good performance [22, 39, 43, 93, 106, 107]. Fuzzy controller is applicable to linear and nonlinear applications [86]. The trial-and-error method is usually used to find fuzzy control rules and membership functions [103]. Recently, scientist have tried to have enhanced performance and increased robustness of FLC's by using neural networks and genetic algorithms in designing such controllers [93, 103, 106].

2.4.3 Hybrid Fuzzy Plus PID Controller

The conventional PID controller is the most popular controller used throughout the industry today because of simple structures and ease of implementation [42, 107–110]. The PID controllers are designed for linear systems and provide a preferable cost, but the presences of nonlinear effect limit their performances [33, 110]. The PID controller is insensitive to parameter changes and exhibits less precision in terms of control response. Fuzzy controllers as an intelligent control method can overcome some shortcomings of the traditional PID because of their knowledge-based nonlinear structural characteristics [109, 111]. However, the disadvantage of FLC produces oscillations in the process response [105]. Hybridization of these two controller structures comes to mind to exploit the beneficial sides of both categories [111]. Otherwise, the trend nowadays is to implement hybrid systems consisting of more than one type of control technique to design a controller that is robust against parameter variations and lead to better performances [112]. Various hybrid controller structures have arisen in literature for various applications such as to control steam distillation process for essential oil extraction system [113–115]; position control of a slider cranks manipulator [109], and control manipulator and stoker-fired boiler show the effectiveness and robustness of the hybrid fuzzy [116].

Mainly, three types known as PI, PD, and PID of FLC's are considered for process control applications [106]. Fuzzy PI type is most common and practical followed by the fuzzy PD [93]. The performance of fuzzy PI is quite satisfactory for linear first-order systems. However, for higher-order systems, systems with large dead time, and for nonlinear systems, fuzzy PI be very poor due to large overshoot and excessive oscillation at a transient state [93, 106]. Fuzzy PI reduces steady states error but yields penalized rise time and settling time. Fuzzy PD can reliably predict and correct large overshoot but do not improve the steady-state response. Fuzzy PD has good performance at the transient state [105]. Fuzzy PD is suitable for a limited class of systems, and they are not recommended in the presence of measurement noise and sudden load disturbance [93, 106]. Therefore, the enhancement of fuzzy PID is expected to gain best performance. The hybrid fuzzy PID (HFPID) controller provides best system responses in terms of transient and steady-state performances when compared to the pure conventional PID or the pure fuzzy controller [111]. Hybrid fuzzy PID achieves a better, consistent and robust controller compared to PID when considering the test on the track the set-point and load disturbance [113]. Even with fixed-structure parameters, an HFPID controller can still have a wide range of control performances [108].

2.4.4 Self-tuning Fuzzy PID Controller

The self-tuning fuzzy PID controllers (STFPID) were designed with the intention to improve the PID controller performances [22, 39, 117]. The self-tuning fuzzy PID controller is the combination of PID controller and fuzzy controllers and act as an effective tool for many control systems [118]. There is a large volume of published studies describing the implementation of STFPID controllers such in steam temperature regulation [40, 119, 120], position control of shape memory alloy actuators [118], industrial hydraulic actuators [117, 121, 122], gasoline refinery catalytic reformer [122] and plastic injection molding process [123]. The STFPID controllers are more attractive for industrial use since can overcome the limitation of PID controllers such as parameters change and disturbed by unknown facts [121, 124].

The STFPID controller applied the fuzzy tuner to tune automatically the parameters of PID controller, which are K_P, K_I, and K_D based on the information error [117, 119, 121]. There are two inputs to the fuzzy inference: the error, $e(t)$ and derivation of error de(t), and three outputs for each respective PID's parameter namely as $K_P^{'}$, $K_I^{'}$, and $K_D^{'}$ [117, 118, 121]. The $e(t)$ is the difference between desired set point and actual output [117]. The ranges of the PID parameters were set as $[K_{P\,min}, K_{P\,max}]$, $[K_{I\,min}, K_{I\,max}]$, and$[K_{D\,min}, K_{D\,max}]$ [117–119, 121]. The ranges were obtained based on experiment using simulation of PID controller [117, 121].

Previous studies have reported that the fuzzy rules, and the numbers of membership functions were developed based on the characteristic for the system, intuition or practical experience and type of the controller [117–121]. The membership functions are defined over the range of input and output variables known as the universe of discourse [120]. Numerous and growing literatures show that the triangular-shape membership function was popularly used to present the linguistic rules [117–121, 124]. Input and output variables membership functions are symmetric to the origin. Generally, the IF-THEN rules that consist of two antecedences and one consequence were applied during a computation algorithm [117, 121]. The max-min fuzzy inference was used in decision-making output whereas the center of gravity (COG) method was used during defuzzification stage [118, 120, 121]. Defuzzification will generate the crisp output that will be used as the control signal to the plant [120, 121].

The performance of the STFPID controller was evaluated using rise time, settling time, percentage overshoots and steady-state error [117, 121]. Much published research has stressed that the STFPID controller is more attractive for industrial used as it offers superior performance compared with the PID controller [121]. The STFPID controller has the advantageous over PID controller by significantly reduce rise time and the percentage overshoot [119, 121]. The experimental result showed that the STFPID controller could achieve good tracking set point regulation and could compensate the disturbances [118].

2.4.5 Fuzzy Logic Controller Design

Figure 2.2 shows the architecture of an FLC which comprises three principal components, which are fuzzification, interface rule and defuzzification [90, 101, 104, 125].

In the FLCs design, the main control parameters of fuzzy mechanism which are scaling factor, membership function and control rules must be tuned properly to reach the optimal performance [126]. There are many methods available through the literature to tune these parameters such as trial-and-error, genetic algorithm, gradient descent method and neural networks [93, 103, 108, 127]. There is still no standard method to design and tuning the scaling factor, the number of rules, input's space partitions and membership functions [109].

The first step in the design of the FLC is to determine the input of the FLCs. Selection of FLCs input is crucial jobs because the input must be significant to maintain the controller response at the desired set-point. Many publications on FLCs used error and change of error as an input because this input is sufficient to give a good performance for control applications [43, 86, 97, 99, 109, 128, 129].

2.4.5.1 Fuzzification and Membership Function

Fuzzification is a process of converting the crisp inputs into the linguistic variables based on the membership function to obtain the confidence values [101, 125, 130]. Fuzzification is necessary during an earlier stage and plays an important role in dealing with uncertain information, which might be objective or subjective in nature [101].

Selection of membership function that are used in fuzzification is important because having a significant effect in the final performance of the controller [131]. Several types of membership functions used in design FLCs such as trapezoidal function, triangular function, bell function, square function, and Gaussian function [43, 101]. Such functions are used in FLC because they lead themselves to manipulation through the use of fuzzy arithmetic [101]. Trapezoidal and triangular functions

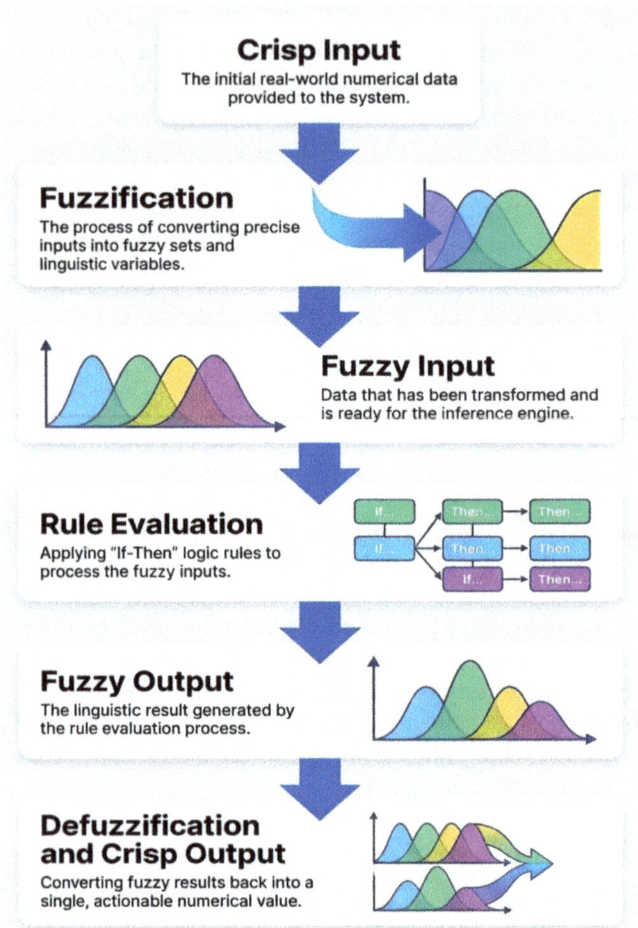

Fig. 2.2 Architecture of fuzzy logic controller

are most used because of simplicity, less memory consumption, fast computation and efficient implementation in steam temperature [43, 126]. Numerous studies used triangular and trapezoidal function in their work and showed the successful results [99, 104, 106, 108, 111, 125].

2.4.5.2 Interfaces Rules

The interface mechanism consists of a set of linguistic control rules and employment rules to infer fuzzy control actions in response to fuzzified inputs [90]. The creation of the rule base is built on general observations and knowledge of problems and usually straightforward to design [108, 125]. These linguistic rules express the relationship between input variables and output variables [105]. Implementing the right rule is important to make sure the stability to the system. The expert knowledge usually in the form of IF-THEN rules, which are easily implemented, and it provides efficient computational [89, 132–134].

2.4.5.3 Defuzzification

Defuzzification is a process which converts the result reaches from the fuzzy set into corresponding universes of discourse output membership function [109, 125, 130]. Some factors need to be considered in selecting the defuzzification methods such as application, the precision of the output and computational time [135]. Popular defuzzification methods are such as center of gravity, center of maximum and height defuzzification method [97, 103, 105, 109, 112, 125, 128–131]. Among them, center of gravity defuzzification method is most common in fuzzy control problems because of gives a precise output evaluation and fast computation time [109, 126]. The center of gravity method will find the balance point of the solution fuzzy region by calculating the weighted mean of the output fuzzy region [105].

2.4.5.4 Scaling Factors

In the fuzzy controller, the scaling factors play a similar role of the gain coefficients of a conventional controller. From the scaling factors, the controller input and output values are mapped into the universe of discourse of the fuzzy set definitions [126]. The scaling factors should be tuned appropriately because higher values of scaling factors will result in faster response but less flexibility of the controller response with changing behavior during the process [90]. To make the fuzzy controller achieve the prospective target, it is necessary to adjust membership functions, control rules and scaling factors [105, 128]. Among them, the control rules and scaling factors play more important roles [105]. Scaling factors can achieve the requisite of real time control [128] and have the highest priority due to their global effect of the control performance [93]. However, there is still no standard method in tuning the scaling factors [105, 109]. Tuning the input and output scaling factors are crucial jobs which in many cases are done through trial and error or based on some training data [93]. In some literatures, they tune the input and output scaling factors by using neural network, self-tuning, genetic algorithm, real-time tuning, switching type and gain scheduling [86, 93, 105].

2.5 Robustness Test and Controller Performance Evaluation

It is very important to consider the robustness test as well as performance in control system design [18]. Generally, robustness test is applied to provide the controller that can produce a smooth control response and also robust to any changes of the operation conditions during running process [113]. Each process is tested on a robustness test which is set-point changes and load disturbance [93]. The objective of the set-point changes is to force the output response of the proposed controller to follow the new set-point either increase or decrease the new set-point by minimizing the error between the reference and actual signal [113]. Meanwhile, the aim of the load disturbance test is to test the ability of the controller in comprising an abrupt disturbance during running process [113]. Many types of disturbances can be applied such as the increase of the water flow rate through the distillation column that can affect the process [113], a combination of white noise and sine-wave noise [109] and changes to the composition of raw materials [129]. Numerous works applied the robustness test to check the capability and performance of their proposed controller for the process [52, 58, 93, 96, 135, 136].

The good controller is defined as a controller that could minimize the error between desired target and actual output in the steady-state manner [86]. The standard performance indexes used in the system's step response are percentage overshoot, rise time, settling time, and accumulated error [103, 138]. The percentage overshoot is the percentage of the maximum excess value versus the final value. The rise time is the time taken for the step response to rise from 10 to 90% of its final value. The accumulated error is the summation of the absolute error of the step response at the sampling instants [103]. The settling time is the time taken to attain the nearly constant value of set-point. This performance index was applied to many applications [93, 94, 100], and it is desirable to optimize a few different system performances to meet some requirement, which is smaller overshoot, minimizing the settling time of the output, fast rise time, smaller steady-state error, and the smaller load effect [97, 105].

References

1. Wang, L., Sun, D.W.: Recent developments in numerical modelling of heating and cooling processes in the food industry—a review. Trends Food Sci. Technol. **14**(10), 408–423 (2003). https://doi.org/10.1016/S0924-2244(03)00151-1
2. Chook, K.C., Tan, A.H.: Identification of an electric resistance furnace. IEEE Trans. Instrum. Meas. **56**(6), 2235–2242 (2007). https://doi.org/10.1109/TIM.2007.907960
3. Masango, P.: Cleaner production of essential oils by steam distillation. J. Clean. Prod. **13**(8), 833–839 (2005). https://doi.org/10.1016/j.jclepro.2004.02.039
4. Pourmortazavi, S.M., Hajimirsadeghi, S.S.: Supercritical fluid extraction in plant essential and volatile oil analysis. J. Chromatograph. A. **1163**(1–2), 2–24 (2007). https://doi.org/10.1016/j.chroma.2007.06.021

5. Wang, L., Weller, C.L.: Recent advances in extraction of nutraceuticals from plants. Trends Food Sci. Technol. **17**(6), 300–312 (2006)
6. Eikani, M.H., Golmohammad, F., Rowshanzamir, S.: Subcritical water extraction of essential oils from coriander seeds (*Coriandrum sativum* L.). J. Food Eng. **80**(2), 735–740 (2007). https://doi.org/10.1016/j.jfoodeng.2006.05.015
7. Sohlberg, B.: Grey box modelling for model predictive control of a heating process. J. Process Control. **13**(3), 225–238 (2003). https://doi.org/10.1016/S0959-1524(02)00030-6
8. Ramchandran, S., Rhinehart, R.R.: A very simple structure for neural network control of distillation. J. Process Control. **5**(2), 115–128 (1995). https://doi.org/10.1016/0959-1524(95)90348-I
9. Mechaqrane, A., Zouak, M.: A comparison of linear and neural network ARX models applied to a prediction of the indoor temperature of a building. Neural Comput. Appl. **13**(1), 20–30 (2004). https://doi.org/10.1007/s00521-004-0401-8
10. Alpbaz, M., Hapoğlu, H., Güresinli, Ç.: Application of nonlinear generalised minimum variance control to a tubular flow reactor. Comput. Chem. Eng. **22**(Suppl.1), 161–166 (1998). https://doi.org/10.1016/s0098-1354(98)00161-6
11. Macmurray, J.C., Himmelblau, D.M.: Modeling and control of a packed distillation column using artificial neural networks. Comput. Chem. Eng. **19**(10), 98–105 (1995). https://doi.org/10.1016/0098-1354(94)00098-9
12. Patil, S.L., Tantau, H.J., Salokhe, V.M.: Modelling of tropical greenhouse temperature by auto regressive and neural network models. Biosyst. Eng. **99**(3), 9–15 (2008). https://doi.org/10.1016/j.biosystemseng.2007.11.009
13. Rahiman, M.H.F., Taib, M.N., Salleh, Y.M.: System identification for essential oil extraction system: an overview. In: Proceedings of 2009 5th International Colloquium on Signal Processing and Its Applications (CSPA 2009) (2009). https://doi.org/10.1109/CSPA.2009.5069252
14. Rahiman, M.H.F., Taib, M.N., Salleh, Y.M.: Black box modeling of steam temperature. In: SCOReD 2006—Proceedings of 2006 4th Student Conference on Research and Development (2006). https://doi.org/10.1109/SCORED.2006.4339334
15. Balakrishnan, V.: System identification: theory for the user (second edition). Automatica. **38**(2), 214–219 (2002). https://doi.org/10.1016/s0005-1098(01)00214-x
16. Frattini Fileti, A.M., Cruz, S.L., Pereira, J.A.F.R.: Control strategies analysis for a batch distillation column with experimental testing. Chem. Eng. Process. Process Intensif. **39**(2), 59–64 (2000). https://doi.org/10.1016/S0255-2701(99)00059-8
17. Karacan, S.: Application of a non-linear long range predictive control to a packed distillation column. Chem. Eng. Process. Process Intensif. **42**(12), 190–195 (2003). https://doi.org/10.1016/S0255-2701(02)00190-3
18. Bala, B.K.: Process dynamics and control. In: Agro-product Processing Technology. CRC Press, Boca Raton (2020). https://doi.org/10.1201/9780429487507-14
19. Slätteke, O., Åström, K.J.: Modeling of a steam heated rotating cylinder—a grey-box approach. Proceedings of the American Control Conference. (2005). https://doi.org/10.1109/acc.2005.1470169
20. Yusoff, Z.M., Muhammad, Z., Rahiman, M.H.F., Taib, M.N.: ARX modeling for down-flowing steam distillation system. In: Proceedings of the 2012 IEEE 8th International Colloquium on Signal Processing and Its Applications (CSPA 2012) (2012). https://doi.org/10.1109/CSPA.2012.6194774
21. Rodriguez Vasquez, J.R., Rivas Perez, R., Sotomayor Moriano, J., Peran Gonzalez, J.R.: System identification of steam pressure in a fire-tube boiler. Comput. Chem. Eng. **32**(12), 2839–2848 (2008). https://doi.org/10.1016/j.compchemeng.2008.01.010
22. Yusoff, Z.M., Muhammad, Z., Rahiman, M.H.F., Taib, M.N.: Real time steam temperature regulation using self-tuning fuzzy pid controller on hydro-diffusion essential oil extraction system. Int. J. Smart Sensing Intell. Syst. **6**(5), 2055–2074 (2013). https://doi.org/10.21307/ijssis-2017-626

23. Yusoff, Z.M.: Self-tuning Fuzzy PID Control of Hydro-diffusion Essential Oil Extraction System. Universiti Teknologi MARA, Malaysia (2014)
24. Ríos-Moreno, G.J., Trejo-Perea, M., Castañeda-Miranda, R., Hernández-Guzmán, V.M., Herrera-Ruiz, G.: Modelling temperature in intelligent buildings by means of autoregressive models. Autom. Constr. **16**(5) (2007). https://doi.org/10.1016/j.autcon.2006.11.003
25. Sohlberg, B.: Hybrid grey box modelling of a pickling process. Control Eng. Pract. **13**(9), 1093–1102 (2005). https://doi.org/10.1016/j.conengprac.2004.11.005
26. Ljung, L., Vicino, A.: Guest editorial: special issue on system identification. IEEE Trans. Autom. Control **50**(10), 1473–(2005). https://doi.org/10.1109/TAC.2005.856638
27. Shojaeefar, M.H., Goudarzi, K., Noorpoor, A.R., Fazelpour, M.: A study of thermal contact using nonlinear system identification models. Am. J. Eng. Applied Sci. **1**(1), 16–23 (2008). https://doi.org/10.3844/ajeassp.2008.16.23
28. Arkov, V., et al.: System identification strategies applied to aircraft gas turbine engines. Annu. Rev. Control. **24**, 67–81 (2000). https://doi.org/10.1016/S1367-5788(00)00004-3
29. Ahmad, S.M., Shaheed, M.H., Chipperfield, A.J., Tokhi, M.O.: Nonlinear modelling of a twin rotor MIMO system using radial basis function networks. In: National Aerospace and Electronics Conference, Proceedings of the IEEE (2000). https://doi.org/10.1109/naecon.2000.894926
30. Nooraii, A., Romagnoli, J.A., Figueroa, J.: Process identification, uncertainty characterisation and robustness analysis of a pilot-scale distillation column. J. Process Control 9(3), 247–264. https://doi.org/10.1016/S0959-1524(98)00042-0
31. Åström, K.J., Bell, R.D.: Drum-boiler dynamics. Automatica. **36**(3), 363–378 (2000). https://doi.org/10.1016/S0005-1098(99)00171-5
32. Muhammad, Z., Yusoff, Z.M., Rahiman, M.H.F., Taib, M.N.: Steam temperature control for steam distillation pot using model predictive control. In: Proceedings of the 2012 IEEE 8th International Colloquium on Signal Processing and Its Applications (CSPA 2012) (2012). https://doi.org/10.1109/CSPA.2012.6194772
33. Muhammad, Z., Yusoff, Z.M., Rahiman, M.H.F., Taib, M.N.: Design and implementation of fuzzy logic controller for induction based steam distillation system. In: Proceedings of the IEEE 10th International Colloquium on Signal Processing and Its Applications (CSPA 2014) (2014). https://doi.org/10.1109/CSPA.2014.6805741
34. Ashar, N.D.B.K., Yusoff, Z.M., Ismail, N., Hairuddin, M.A.: ARX model identification for the real-time temperature process with Matlab-arduino implementation. ICIC Express Lett. **14**(2), 167–174 (2020). https://doi.org/10.24507/icicel.14.02.103
35. Muhammad, Z., Yusoff, Z.M., Rahiman, M.H.F., Taib, M.N.: Modeling of steam distillation pot with ARX model. In: Proceedings of the IEEE 8th International Colloquium on Signal Processing and Its Applications (CSPA 2012) (2012). https://doi.org/10.1109/CSPA.2012.6194717
36. Karacan, S., Hapoğlu, H., Alpbaz, M.: Application of optimal adaptive generalized predictive control to a packed distillation column. Chem. Eng. J. **84**(3), 539–548 (2001). https://doi.org/10.1016/S1385-8947(01)00130-9
37. Shaw, A.M., Doyle, F.J.: Multivariable nonlinear control applications for a high purity distillation column using a recurrent dynamic neuron model. J. Process Control. **7**(4), 255–268 (1997). https://doi.org/10.1016/S0959-1524(97)00002-4
38. Tajjudin, N., Yusuf, Z., Rahiman, M.H.F., Taib, M.N.: Self tuning PID controller for steam distillation essential oil extraction system. In: Proceedings of the 6th International Colloquium on Signal Processing and Its Applications (CSPA 2010) (2010). https://doi.org/10.1109/CSPA.2010.5545292
39. Muhammad, Z., Jamil, K.A., Yusoff, Z.M., Rahiman, M.H.F.: Application of self-tuning fuzzy PID (STFPID) controller on industrial essential oil extraction system using system identification approach. J. Telecommun. Electron. Comput. Eng. **10**(1–3), 77–81 (2018)
40. Isomursu, P., Rauma, T.: Self-tuning fuzzy logic controller for temperature control of superheated steam. In: IEEE International Conference on Fuzzy Systems, pp. 1404–1408 (1994). https://doi.org/10.1109/fuzzy.1994.343927

41. Kasuan, N., Yusoff, Z.M., Muhammad, Z., Nordin, M.N.N., Rahiman, M.H.F., Taib, M.N.: Essential oil extraction with automated steam distillation: FMRLC for steam temperature regulation. In: Proceedings of the IEEE International Conference on Control System, Computing and Engineering (ICCSCE 2012) (2012). https://doi.org/10.1109/ICCSCE.2012.6487141
42. Yusoff, Z.M., Muhammad, Z., Taib, M.N., Rahiman, M.H.F.: Implementation of hybrid fuzzy plus PID controller in real time steam temperature control. In: Proceedings of the 2013 IEEE 3rd International Conference on System Engineering and Technology (ICSET 2013) (2013). https://doi.org/10.1109/ICSEngT.2013.6650199
43. Mohammad, N.N., Kasuan, N., Rahiman, M.H.F., Taib, M.N.: Steam temperature control using fuzzy logic for steam distillation essential oil extraction process. In: Proceedings of the 2011 IEEE Control and System Graduate Research Colloquium (ICSGRC 2011) (2011). https://doi.org/10.1109/ICSGRC.2011.5991829
44. Mahabob, N.Z., et al.: A study on ANN performance towards three significant compounds of high quality agarwood oil. In: 2022 IEEE 18th International Colloquium on Signal Processing and Applications, CSPA 2022—Proceeding, pp. 116–120. Institute of Electrical and Electronics Engineers Inc. (2022). https://doi.org/10.1109/CSPA55076.2022.9782017
45. Bisgin, H., et al.: Comparing SVM and ANN based machine learning methods for species identification of food contaminating beetles. Sci. Rep. **8**, 1–12 (2018). https://doi.org/10.1038/s41598-018-24926-7
46. Mahabob, N.Z., Amidon, A.F.M., Yusoff, Z.M., Ismail, N., Tajjudin, S.N.: Comparison of ANN performance towards agarwood oil compounds pre-processing based on principal component analysis (PCA) and stepwise regression selection method. J. Electr. Electron Syst. Res. **19**, 45–51 (2021). https://doi.org/10.24191/jeesr.v19i1.007
47. Häggblom, K.E.: Combined internal model and inferential control of a distillation column via closed-loop identification. J. Process Control. **6**, 223–231 (1996). https://doi.org/10.1016/0959-1524(95)00032-1
48. Häggblom, K.E., Böling, J.M.: Multimodel identification for control of an ill-conditioned distillation column. J. Process Control. **8**, 187–195 (1998). https://doi.org/10.1016/S0959-1524(97)00040-1
49. Safavi, A.A., Nooraii, A., Romagnoli, J.A.: A hybrid model formulation for a distillation column and the on-line optimisation study. Chem. Eng. J. **72**, 43–52 (1999). https://doi.org/10.1016/S0959-1524(98)00041-9
50. Perspectives on system identification: LLjung, L. Annu. Rev. Control. **34**, 1–12 (2010). https://doi.org/10.1016/j.arcontrol.2009.12.001
51. Ljung, L.: System identification toolbox. In: The Matlab User's Guide, vol. 1 (2011)
52. Tsai, C.C., Huang, C.H.: Model reference adaptive predictive control for a variable-frequency oil-cooling machine. IEEE Trans. Indus. Electron. **51**(2), 411–419 (2004). https://doi.org/10.1109/TIE.2004.825229
53. Liu, M., Wang, Q.G., Huang, B., Hang, C.C.: Improved identification of continuous-time delay processes from piecewise step tests. J. Process Control. **17**(1), 51–57 (2007). https://doi.org/10.1016/j.jprocont.2006.08.002
54. Hapoglu, H., Karacan, S., Erten Koca, Z.S., Alpbaz, M.: Parametric and nonparametric model based control of a packed distillation column. Chem. Eng. Process. Process Intensif. **40**(6), 555–564 (2001). https://doi.org/10.1016/S0255-2701(00)00152-5
55. Maiti, S.N., Saraf, D.N.: Adaptive dynamic matrix control of a distillation column with closed-loop online identification. J. Process Control. **5**(5), 315–327 (1995). https://doi.org/10.1016/0959-1524(95)00002-8
56. Gopaluni, R.B., Patwardhan, R.S., Shah, S.L.: The nature of data pre-filters in MPC relevant identification—open- and closed-loop issues. Automatica. **39**(9), 1657–1662 (2003). https://doi.org/10.1016/S0005-1098(03)00146-8
57. Scheers, B., Le Nir, V.: Pseudo-random binary sequence selection for delay and add direct sequence spread spectrum modulation scheme. IEEE Commun. Lett. **14**(11), 1002–1004 (2010). https://doi.org/10.1109/LCOMM.2010.100410.100906

58. Kasuan, N., Taib, M.N., Fazalul Rahiman, M.H.: Model reference adaptive controller to regulate steam temperature in distillation process for essential oil extraction. In: Proceedings of the 2011 IEEE 7th International Colloquium on Signal Processing and Its Applications (CSPA 2011), Penang, Malaysia (2011). https://doi.org/10.1109/CSPA.2011.5759891
59. Li, Y., Jun, Z., Qixin, Q.: An improved pseudo-random binary sequence design for multivariable system identification. In: Proceedings of the World Congress on Intelligent Control and Automation (WCICA) (2006). https://doi.org/10.1109/WCICA.2006.1712657
60. Fernholz, G., Foechterle, C., Engell, S., Sonnenschein, M.: Controller design for the heating system of a semibatch distillation column. In: IEEE Conference on Control Applications—Proceedings (1998). https://doi.org/10.1109/cca.1998.728235
61. Karacan, S., Hapoglu, H., Alpbaz, M.: Generalized predictive control to a packed distillation column for regulatory problems. Comput. Chem. Eng. **22**(SUPPL. 1), S629–S632 (1998). https://doi.org/10.1016/s0098-1354(98)00111-2
62. Ahmed, S., Huang, B., Shah, S.L.: Identification from step responses with transient initial conditions. J. Process Control. **18**(2), 163–172 (2008). https://doi.org/10.1016/j.jprocont.2007.07.009
63. Alpbaz, M., Karacan, S., Cabbar, Y., Hapoğlu, H.: Application of model predictive control and dynamic analysis to a pilot distillation column and experimental verification. Chem. Eng. J. **88**(1–3), 163–174 (2002). https://doi.org/10.1016/S1385-8947(01)00304-7
64. Ying, C.M., Joseph, B.: Identification of linear systems using polynomial kernels in the frequency domain. J. Process Control. **12**(5), 617–627 (2002). https://doi.org/10.1016/S0959-1524(01)00053-1
65. Yusoff, Z.M., Muhammad, Z., Rahiman, M.H.F., Taib, M.N.: PRBS perturbation and modeling of hydro-diffusion steam distillation essential oil extraction system under feed-in disturbance. In: Proceedings of the 2012 IEEE 8th International Colloquium on Signal Processing and Its Applications (CSPA 2012) (2012). https://doi.org/10.1109/CSPA.2012.6194773
66. Rahiman, M.H.F., Taib, M.N., Salleh, Y.M.: Selection of training data for modeling essential oil extraction system using NNARX structure. In: ICCAS 2007—International Conference on Control, Automation and Systems (2007). https://doi.org/10.1109/ICCAS.2007.4406607
67. Goodwin, G.C., Agüero, J.C., Welsh, J.S., Yuz, J.I., Adams, G.J., Rojas, C.R.: Robust identification of process models from plant data. J. Process Control. **18**(9), 821–834 (2008). https://doi.org/10.1016/j.jprocont.2008.06.004
68. Norton, J.P.: System identification and control design using P.I.M. + software. I. D. Landau. Automatica. **27**(6), 1056–1057 (1991). https://doi.org/10.1016/0005-1098(91)90148-U
69. Rahiman, M.H.F., Taib, M.N., Salleh, Y.M.: Assessment of NNARX structure as a global model for self-refilling steam distillation essential oil extraction system. In: Proceedings of the International Symposium on Information Technology 2008 (ITSim 2008) (2008). https://doi.org/10.1109/ITSIM.2008.4632020
70. Ohata, A., Furuta, K., Nita, H.: Identification of nonlinear ARX model with input and output dependent coefficients. In: Proceedings of the IEEE International Conference on Control Applications (2006). https://doi.org/10.1109/CACSD-CCA-ISIC.2006.4777044
71. Raghavan, H., Tangirala, A.K., Gopaluni, R.B., Shah, S.L.: Identification of chemical processes with irregular output sampling. Control Eng. Pract. **14**(5), 467–480 (2006). https://doi.org/10.1016/j.conengprac.2005.01.015
72. Cott, B.J.: Summary of the process identification workshop at the 1992 Canadian chemical engineering conference. J. Process Control. **5**(2), 125–128 (1995). https://doi.org/10.1016/0959-1524(95)90347-H
73. Pearson, R.K.: Nonlinear input/output modelling. J. Process Control. **5**(4), 211–228 (1995). https://doi.org/10.1016/0959-1524(95)00014-H
74. Tufa, L.D., Ramasamy, M., Patwardhan, S.C., Shuhaimi, M.: Development of Box-Jenkins type time series models by combining conventional and orthonormal basis filter approaches. J. Process Control. **20**(1), 108–120 (2010). https://doi.org/10.1016/j.jprocont.2009.07.009
75. Rahiman, M.H.F., Taib, M.N., Salleh, Y.M.: Performance of multi-step-ahead-prediction ARX for steam temperature in a self-refilling distillation essential oil extraction system. In: ICCAS

2007—International Conference on Control, Automation and Systems (2007). https://doi.org/10.1109/ICCAS.2007.4406608
76. Ismail, N., Tajjudin, N., Rahiman, M.H.F., Taib, M.N.: Modeling of dynamic response of essential oil extraction process. In: Proceedings of 2009 5th International Colloquium on Signal Processing and Its Applications (CSPA 2009) (2009). https://doi.org/10.1109/CSPA.2009.5069238
77. Ismail, N., Rahiman, M.H.F., Taib, M.N.: Noise variation test for ARX model of a heating process in essential oil extraction plant. In: Proceedings of the CSPA 2010: 2010 6th International Colloquium on Signal Processing and Its Applications (2010). https://doi.org/10.1109/CSPA.2010.5545304
78. Ismail, N., Rahiman, M.H.F., Taib, M.N.: Investigation of ARX model on partial input-output data in heating process. In: Proceedings of the 2011 IEEE Control and System Graduate Research Colloquium (ICSGRC 2011) (2011). https://doi.org/10.1109/ICSGRC.2011.5991820
79. Tsai, P.F., Chu, J.Z., Jang, S.S., Shieh, S.S.: Developing a robust model predictive control architecture through regional knowledge analysis of artificial neural networks. J. Process Control. **13**(5), 411–424 (2003). https://doi.org/10.1016/S0959-1524(02)00067-7
80. Åström, K.J., Hägglund, T.: Revisiting the Ziegler-Nichols step response method for PID control. J. Process Control. **14**(6), 635–650 (2004). https://doi.org/10.1016/j.jprocont.2004.01.002
81. Kansha, Y., Jia, L., Chiu, M.S.: Self-tuning PID controllers based on the Lyapunov approach. Chem. Eng. Sci. **63**(10), 2697–2705 (2008). https://doi.org/10.1016/j.ces.2008.02.026
82. Hang, C.C., Sin, K.K.: On-line auto tuning of PID controllers based on the cross-correlation technique. IEEE Trans. Ind. Electron. **38**(6), 422–428 (1991). https://doi.org/10.1109/41.107098
83. Shen, J.C.: New tuning method for PID controller. ISA Trans. **41**(4), 473–484 (2002). https://doi.org/10.1016/s0019-0578(07)60103-7
84. Lin, F., Brandt, R.D., Saikalis, G.: Self-tuning of PID controllers by adaptive interaction. In: Proceedings of the American Control Conference, vol. 5 (2000). https://doi.org/10.1109/acc.2000.879256
85. Tajjudin, M., Ishak, N., Ismail, H., Rahiman, M.H.F., Adnan, R.: Optimized PID control using Nelder-Mead method for electro-hydraulic actuator systems. In: Proceedings of the 2011 IEEE Control and System Graduate Research Colloquium (ICSGRC 2011) (2011). https://doi.org/10.1109/ICSGRC.2011.5991836
86. Yusuf, Z., Janin, Z., Taib, M.N.: Self Tuning Fuzzy Logic controller with Smith structure for controlling glycerin bleaching temperature. In: Proceedings of the ICSGRC 2010: 2010 IEEE Control and System Graduate Research Colloquium (2010). https://doi.org/10.1109/ICSGRC.2010.5562526
87. Ishak, A.A., Ahmad, A.: Application of numerical technique in tangent analysis of a discrete step response data. In: 2002 Student Conference on Research and Development: Globalizing Research and Development in Electrical and Electronics Engineering (SCOReD 2002)—Proceedings (2002). https://doi.org/10.1109/SCORED.2002.1033045
88. Gaing, Z.L.: A particle swarm optimization approach for optimum design of PID controller in AVR system. IEEE Trans. Energy Convers. **19**(2), 384–391 (2004). https://doi.org/10.1109/TEC.2003.821821
89. Hsu, C.F., Lee, B.K.: FPGA-based adaptive PID control of a DC motor driver via sliding-mode approach. Expert Syst. Appl. **38**(9), 11830–11839 (2011). https://doi.org/10.1016/j.eswa.2011.02.185
90. Yusuf, Z., Janin, Z., Taib, M.N.: Fuzzy Logic controller for glycerin bleaching temperature control. In: Proceedings of the CSPA 2010: 2010 6th International Colloquium on Signal Processing and Its Applications (2010). https://doi.org/10.1109/CSPA.2010.5545295
91. Visioli, A.: Tuning of PID controllers with fuzzy logic. IEE Proc. Control Theory Appl. **148**(1), 1–8 (2001). https://doi.org/10.1049/ip-cta:20010232

92. Muhammad, Z., Yusoff, Z.M., Rahiman, M.H.F., Taib, M.N.: The self-tuning fuzzy PID to regulate the steam temperature in real-time process. In: Proceedings of the 2015 IEEE Conference on System, Process and Control (ICSPC 2015) (2016). https://doi.org/10.1109/SPC.2015.7473554
93. Mudi, R.K., Pal, N.R.: A robust self-tuning scheme for PI- and PD-type fuzzy controllers. IEEE Trans. Fuzzy Syst. **7**(1), 2–16 (1999). https://doi.org/10.1109/91.746295
94. Zhao, Z.Y., Tomizuka, M., Isaka, S.: Fuzzy gain scheduling of PID controllers. IEEE Trans. Syst. Man Cybern. **23**(5), 1392–1405 (1993). https://doi.org/10.1109/21.260670
95. Tavakoli, S., Tavakoli, M.: Optimal tuning of PID controllers for first order plus time delay models using dimensional analysis. In: International Conference on Control and Automation (2003). https://doi.org/10.1109/icca.2003.1595161
96. Golshan, M., MacGregor, J.F., Bruwer, M.J., Mhaskar, P.: Latent variable model predictive control (LV-MPC) for trajectory tracking in batch processes. J. Process Control. **20**(4), 488–499 (2010). https://doi.org/10.1016/j.jprocont.2010.01.007
97. Tang, K.S., Man, K.F., Chen, G., Kwong, S.: An optimal fuzzy PID controller. IEEE Trans. Ind. Electron. **48**(4), 751–759 (2001). https://doi.org/10.1109/41.937407
98. Huzmezan, M., Gough, B., Kovac, S.: Advanced control of batch reactor temperature. In: Proceedings of the American Control Conference, vol. 2 (2002). https://doi.org/10.1109/ACC.2002.1023175
99. Threesinghawong, N., Sanposh, P., Chinthaned, N., Srinophakun, P.: Temperature control of Koji process by using fuzzy PD controller. In: 5th International Conference on Electrical Engineering/Electronics, Computer, Telecommunications and Information Technology, ECTI-CON 2008 (2008). https://doi.org/10.1109/ECTICON.2008.4600521
100. Wu, H., Yu, J.Q.: Study on smith-fuzzy self-tuning pid control algorithm of temperature for regenerative magnesium reduction furnace. In: 2009 Chinese Control and Decision Conference (CCDC 2009) (2009). https://doi.org/10.1109/CCDC.2009.5191587
101. Lee, C.C.: Fuzzy logic in control systems: fuzzy logic controller—Part I. IEEE Trans. Syst. Man Cybern. **20**(2), 404–418 (1990). https://doi.org/10.1109/21.52551
102. Cao, S.G., Rees, N.W., Feng, G.: Analysis and design for a class of complex control systems part II: Fuzzy controller design. Automatica. **33**(6), 1029–1043 (1997). https://doi.org/10.1016/s0005-1098(97)00011-3
103. Zhou, Y.S., Lai, L.Y.: Optimal design for fuzzy controllers by genetic algorithms. IEEE Trans. Ind. Appl. **36**(1), 93–97 (2000). https://doi.org/10.1109/28.821802
104. Xiaodong, Z., Senlin, X., Hongbo, Z.: A new strategy for batch reactor's temperature control. In: 2009 WRI World Congress on Computer Science and Information Engineering, CSIE 2009 (2009). https://doi.org/10.1109/CSIE.2009.579
105. Chung, H.Y., Chen, B.C., Lin, J.J.: A PI-type fuzzy controller with self-tuning scaling factors. Fuzzy Sets Syst. **93**(1), 19–28 (1998). https://doi.org/10.1016/S0165-0114(96)00215-1
106. Mudi, R.K., Pal, N.R.: A self-tuning fuzzy PI controller. Fuzzy Sets Syst. **115**(2), 327–338 (2000). https://doi.org/10.1016/S0165-0114(98)00147-X
107. Yusoff, Z.M., Muhammad, Z., Taib, M.N., Rahiman, M.H.F.: Steam temperature control of hydro-diffusion essential oil extraction system using hybrid-fuzzy plus PID controller. In: Proceedings of the 2014 IEEE Conference on System, Process and Control, ICSPC 2014 (2014). https://doi.org/10.1109/SPC.2014.7086239
108. Er, M.J., Sun, Y.L.: Hybrid fuzzy proportional-integral plus conventional derivative control of linear and nonlinear systems. IEEE Trans. Ind. Electron. **48**(6), 1109–1117 (2001). https://doi.org/10.1109/41.969389
109. Ahn, K.K., Truong, D.Q.: Online tuning fuzzy PID controller using robust extended Kalman filter. J. Process Control. **19**(6), 1011–1023 (2009). https://doi.org/10.1016/j.jprocont.2009.01.005
110. Yusoff, Z.M., Muhammad, Z., Rahiman, M.H.F., Taib, M.N.: Application of hybrid fuzzy plus PID for steam temperature control in hydro-diffusion essential oil extraction system. In: Proceedings of the 2014 IEEE 10th International Colloquium on Signal Processing and Its Applications, CSPA 2014 (2014). https://doi.org/10.1109/CSPA.2014.6805740

111. Erenoglu, I., Eksin, E., Yesil, E., Guzelkaya, M.: An intelligent hybrid fuzzy PID controller. In: 20th European Conference on Modelling and Simulation: Modelling Methodologies and Simulation Key Technologies in Academia and Industry (ECMS 2006) (2006). https://doi.org/10.7148/2006-0062
112. Adnan, R., Tajjudin, M., Ishak, N., Ismail, H., Rahiman, M.H.F.: Self-tuning fuzzy PID controller for electro-hydraulic cylinder. In: Proceedings of the 2011 IEEE 7th International Colloquium on Signal Processing and Its Applications (CSPA 2011) (2011). https://doi.org/10.1109/CSPA.2011.5759909
113. Kasuan, N., Yusuf, Z., Taib, M.N., Rahiman, M.H.F., Tajuddin, N., Aziz, M.A.A.: Robust steam temperature regulation for distillation of essential oil extraction process using hybrid fuzzy-PD plus PID controller. World Acad. Sci. Eng. Technol. **47**, 102–107 (2010)
114. Yusoff, Z.M., Muhammad, Z., Rahiman, M.H.F., Taib, M.N.: Analysing the dynamic behaviour of hydro-diffusion system using step test. In: Proceedings of the 2015 IEEE Conference on System, Process and Control (ICSPC 2015) (2016). https://doi.org/10.1109/SPC.2015.7473552
115. Yusoff, Z.M., Nordin, M.N.N., Rahiman, M.H.F., Adnan, R., Taib, M.N.: Characterization of down-flowing steam distillation system using step test analysis. In: Proceedings of the 2011 IEEE Control and System Graduate Research Colloquium (ICSGRC 2011) (2011). https://doi.org/10.1109/ICSGRC.2011.5991856
116. Li, W.: Design of a hybrid fuzzy logic proportional plus conventional integral-derivative controller. IEEE Trans. Fuzzy Syst. **6**(4), 573–581 (1998). https://doi.org/10.1109/91.728430
117. Rahmat, M.F., Has, Z.: Application of self-tuning fuzzy PID controller on industrial hydraulic actuator using system identification approach. Int. J. Smart Sensing Intell. Syst. **2**(2), 250–270 (2009). https://doi.org/10.21307/ijssis-2017-349
118. Ahn, K.K., Nguyen, B.K.: Position control of shape memory alloy actuators using self tuning fuzzy PID controller. Int. J. Control Autom. Syst. **4**(6), 704–713 (2006)
119. Adnan, R., Tajjudin, M., Ishak, N., Ismail, H., Rahiman, M.H.F., Arshad, N.M.: Comparison between self-tuning Fuzzy-PID and pole-placement PID with application to saturated steam temperature regulation. In: Proceedings of the 2012 International Conference on System Engineering and Technology (ICSET 2012) (2012). https://doi.org/10.1109/ICSEngT.2012.6339283
120. Gao, Z., Trautzsch, A.A., Dawson, J.G.: A stable self-tuning fuzzy logic control system for industrial temperature regulation. IEEE Trans. Indus. Appl. **38**(2), 414–424 (2002). https://doi.org/10.1109/28.993163
121. Sinthipsomboon, K., Hunsacharoonroj, I., Khedari, J., Pongaen, W., Pratumsuwan, P.: A hybrid of fuzzy and fuzzy self-tuning PID controller for servo electro-hydraulic system. In: Proceedings of the 2011 6th IEEE Conference on Industrial Electronics and Applications, ICIEA 2011 (2011). https://doi.org/10.1109/ICIEA.2011.5975583
122. Bare, W.H., Mulholland, R.J., Sofer, S.S.: Design of a self-tuning rule based controller for a gasoline refinery catalytic reformer. IEEE Trans. Automat. Contr. **35**(2), 156–164 (1990). https://doi.org/10.1109/9.45171
123. Tsai, C.C., Lu, C.H.: Multivariable self-tuning temperature control for plastic injection molding process. IEEE Trans. Ind. Appl. **34**(2), 310–318 (1998). https://doi.org/10.1109/28.663473
124. Daugherity, W.C., Rathakrishnan, B., Yen, J.: Performance evaluation of a self-tuning fuzzy controller (1992). https://doi.org/10.1109/fuzzy.1992.258646
125. Carvajal, J., Chen, G., Ogmen, H.: Fuzzy PID controller: design, performance evaluation, and stability analysis. Inf. Sci. (N. Y.). **123**(3–4), 249–278 (2000). https://doi.org/10.1016/S0020-0255(99)00127-9
126. Fileti, A.M.F., Antunes, A.J.B., Silva, F.V., Silveira, V., Pereira, J.A.F.R.: Experimental investigations on fuzzy logic for process control. Control Eng. Pract. **15**(9), 1149–1160 (2007). https://doi.org/10.1016/j.conengprac.2007.01.009
127. Guely, F., Siarry, P.: Gradient descent method for optimizing various fuzzy rule bases. In: 1993 IEEE International Conference on Fuzzy Systems (1993). https://doi.org/10.1109/fuzzy.1993.327570

128. Woo, Z.W., Chung, H.Y., Lin, J.J.: A PID type fuzzy controller with self-tuning scaling factors. Fuzzy Sets Syst. **115**(2), 321–326 (2000). https://doi.org/10.1016/S0165-0114(98)00159-6
129. Moon, U.C., Lee, K.Y.: Hybrid algorithm with fuzzy system and conventional PI control for the temperature control of TV glass furnace. IEEE Trans. Control Syst. Technol. **11**(4), 548–554 (2003). https://doi.org/10.1109/TCST.2003.813385
130. Chen, W.C., Chang, N.B., Chen, J.C.: Rough set-based hybrid fuzzy-neural controller design for industrial wastewater treatment. Water Res. **37**(1), 95–107 (2003). https://doi.org/10.1016/S0043-1354(02)00255-5
131. Eker, I., Torun, Y.: Fuzzy logic control to be conventional method. Energy Convers. Manag. **47**(4), 377–394 (2006). https://doi.org/10.1016/j.enconman.2005.05.008
132. D'Errico, G.E.: Fuzzy control systems with application to machining processes. J. Mater. Process. Technol. **109**(1–2), 38–43 (2001). https://doi.org/10.1016/S0924-0136(00)00773-1
133. Li, H.X., Gatland, H.B.: Conventional fuzzy control and its enhancement. IEEE Trans. Syst. Man Cybern. Part B Cybern. 26(5), 703–712 (1996). https://doi.org/10.1109/3477.537321
134. Berenji, H.R., Khedkar, P.: Learning and tuning fuzzy logic controllers through reinforcements. IEEE Trans. Neural Netw. **3**(5), 724–740 (1992). https://doi.org/10.1109/72.159061
135. Lee, Y., Park, S., Lee, M., Brosilow, C.: PID controller tuning for desired closed-loop responses for SI/SO systems. AIChE J. **44**(1), 106–115 (1998). https://doi.org/10.1002/aic.690440112
136. Cheng, C., Chiu, M.S.: Robust PID controller design for nonlinear processes using JITL technique. Chem. Eng. Sci. **63**(21), 5228–5237 (2008). https://doi.org/10.1016/j.ces.2008.07.013
137. Hägglund, T.: An industrial dead-time compensating PI controller. Control Eng. Pract. **4**(6), 749–756 (1996). https://doi.org/10.1016/0967-0661(96)00065-2
138. Zhang, Y., Fei, Q., Lu, J., Lei, W., Wu, Q.: Performance criteria research on PSO-PID control systems. In: Proceedings of the 2010 International Conference on Intelligent Computing and Cognitive Informatics (ICICCI 2010) (2010). https://doi.org/10.1109/ICICCI.2010.51

Chapter 3
Principles and Concepts

Abstract This chapter presents the fundamental principles and concepts of system identification and controller design relevant to dynamic systems, particularly for temperature control applications. It begins by introducing the role of input signals in system identification, highlighting commonly used signals such as step functions and pseudo-random binary sequences (PRBS), which are essential for capturing system dynamics effectively. The chapter further explains the basic concepts of linear system identification, including input–output relationships, disturbance modeling, and the development of mathematical models using structures such as the Auto-Regressive with eXogenous input (ARX) model. Key procedures in model development, including parameter estimation using least-squares methods and model validation techniques such as one-step-ahead prediction, residual analysis, coefficient of determination, root mean square error, and correlation tests, are also discussed. In addition, the chapter introduces conventional and intelligent control strategies. The structure and operation of the proportional–integral–derivative (PID) controller and its tuning method are described, followed by the limitations of traditional controllers in handling nonlinear and complex systems. To address these challenges, the chapter presents the fundamental concepts of fuzzy logic, including fuzzy sets, membership functions, and rule-based inference. The design and operation of fuzzy logic controllers are explained, along with hybrid approaches such as fuzzy–PID and self-tuning fuzzy PID controllers, which offer improved adaptability and control performance. Finally, the chapter discusses performance evaluation criteria for control systems, including rise time, settling time, and percentage overshoot. Overall, this chapter provides essential theoretical knowledge that serves as a foundation for the development of advanced modeling and control strategies in dynamic systems.

3.1 Input Signal

The input signal used in identification can have a significant influence on the estimated model. There are several aspects need to be considered in selecting of input signal. For frequency analyzing the input must be a sinusoid, for transient analysis a step or

Z. Mohd Yusoff et al., *Fuzzy Logic, Smart Control, and the Future of Essential Oils*,
Studies in Fuzziness and Soft Computing 442,
https://doi.org/10.1007/978-981-92-2084-7_3

an impulse and for correlation analysis a white noise or a pseudo random sequence. The input signal is advisable must be persistently exciting [1]. The aim to use different input signals is to drive out the behavior of the system interest at particular dynamics, which are to be characterized [2].

3.1.1 Step Function

A step function is defined as:

$$u(t) = f(x) = \begin{cases} 0, t < 0 \\ u_0, t \geq 0 \end{cases}, \tag{3.1}$$

where u_0 is the chosen amplitude. The system will be relaxed for $t < 0$ but once $t = 0$, the function will drive the system up to u_0 [3]. Step response is the time behavior of the outputs for a general system when its input changes from zero to one, when its control input is step function. Step responses can very well furnish that information to a sufficient degree of accuracy when the goal is to determine some basic control-related characteristics in the transient analysis such as static gain, dominating time constants and time delay [3, 4]. A plot of the output response to a process by using step change in input also referred as the process reaction curve by performing a tangent line to the steepest point of the response curve, as illustrated in Fig. 3.1 [5–7]. Some well-known rules for tuning simple regulators such as Ziegler–Nichols rule are based on model information reached in step responses [3].
where τ is the time constant, θ is the time delay and K is the process gain. Time constant, τ is the time when the response signal reaches 63% of its final value [8]. Dead time, θ is the delay in the output response until the process variable first begins to respond. Meanwhile, process gain, K is the output change during the process over the step input change. Equation (3.2) defined the process gain [9]:

$$K = \frac{\text{Change in measured process variable } (\Delta \text{PV})}{\text{Change in manipulated variable } (\Delta \text{MV})}. \tag{3.2}$$

3.1.2 Pseudo-Random Binary Sequences

Pseudo-random binary sequences (PRBS) are sequences of rectangular pulses, modulated in width, deterministic signal with white-noise-like properties and have a spectral content rich in frequencies [3, 5, 10]. The term *pseudo random* indicated that the input is characterized by sequence length that has the spectral characteristics of random signal but over a large time horizon, they a periodic [5, 10]. The actual sequence of inputs can be repeated multiple times [5]. The advantage of PRBS is

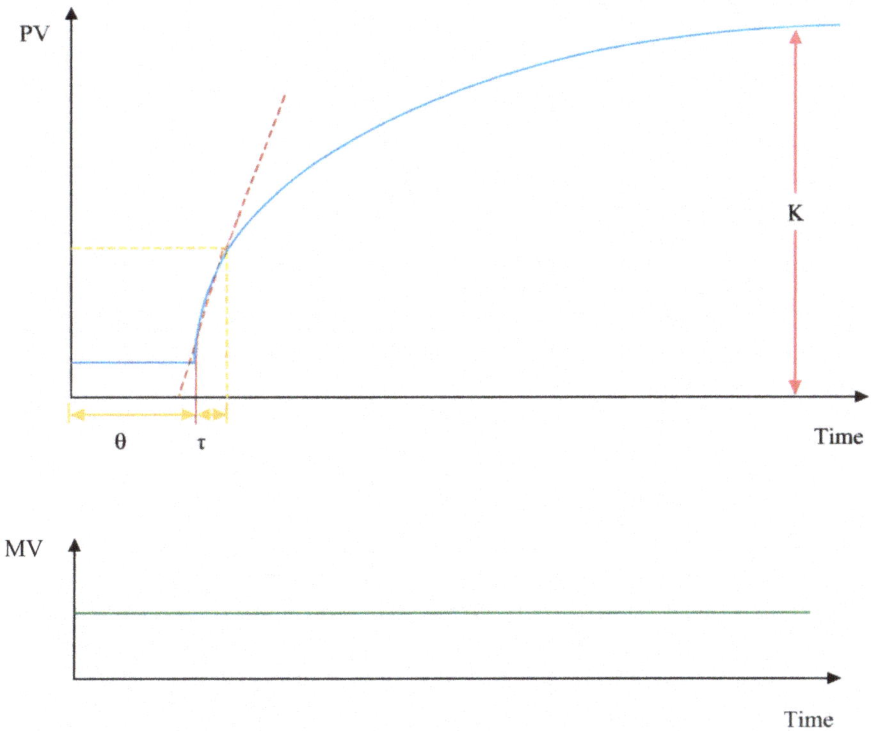

Fig. 3.1 Open loop output

that the input excitation can be concentrated ranges that correspond to the process dynamics and important for control system design. Figure 3.2 shown the PRBS input [5];

The PRBS signal is varied between two levels, which are ± V, and it can switch between one level to another at $1/2(N+1)$ intervals at + V and $1/2(N-1)$ intervals at – V, where N is the odd integer [5]. Two important parameters need to be considering in design the PRBS is the duration of the switching sequence (N_s) and the switching

Fig. 3.2 PRBS inputs

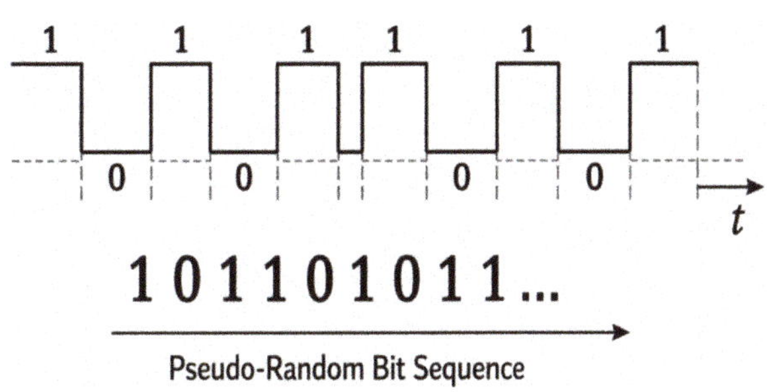

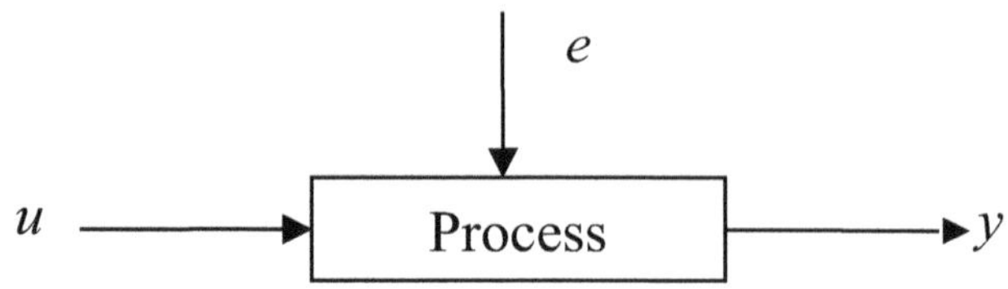

Fig. 3.3 A system block diagram

time or clock period, T_{sw}. Switching time is the minimum time between changes at the level of signal, and it is the integer value multiple with sampling time, Δt. The signal repeats itself after $N_{\text{s}}T_{\text{sw}}$. The actual input sequence sequences PRBS is generated by an n shift registers such that $N_{\text{s}} = 2^{\text{n}-1}$.

3.2 Linear System Identification

Figure 3.3 shows the general block diagram for a dynamic system [2]:

It consists of input, output, and disturbance and these signals are functions of time, where the value is denoted as input, $u(t)$, output, $y(t)$, and error, $e(t)$, respectively. Input signal means external signals that can be manipulated by the observer. Output signal is the observable signals that are of interest. Disturbance is the signals beyond our control that can affect the system and can be divided into directly measured, and only observed through their influence over the output [3, 4, 11]. Input–output data usually collected simultaneously, and signal recorded at discrete time instants, which are typically sampled with a sampling interval of T time units.

3.3 ARX Model Structure

The Auto-regression with exogenous input (ARX) model structure is given by Eq. (3.3) below [3]:

$$A(q)y(t) = B(q)u(t) + e(t), \tag{3.3}$$

where the polynomials $A(q)$ and $B(q)$ are given by

$$A(q) = 1 + a_1q^{-1} + \cdots + a_{n_{\text{a}}}q^{-n_{\text{a}}} \tag{3.4}$$

$$B(q) = 1 + b_0 + b_1q^{-1} + \cdots + b_{n_{\text{a}}}q^{-n_{\text{b}}}, \tag{3.5}$$

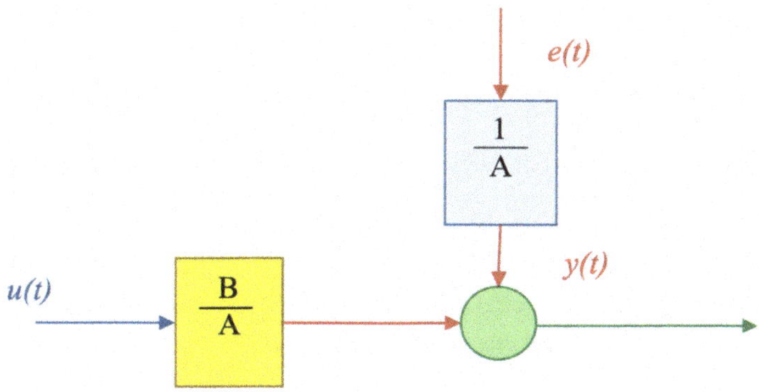

Fig. 3.4 The ARX model structure

where q^{-1} is the backward shift operator. The following block diagram in Fig. 3.4 shows the ARX model structure [2]:

3.4 Model Estimation

The goal of parameter estimation is to estimate unknown model parameters, so that the model fit becomes good in the sense of some criterion. The least-squares approach is a standard method for linear models that minimizes the sum of the squared residuals [5]. The ARX uses the least-squares estimation approach, also known as linear regression, to estimate the parameters of an ARX model. The input–output relationship of ARX model is defined as [3]:

$$y(t) + a_1 y(t-1) + \cdots + a_{n_a} y(t-n_a) = b_1 u(t-1) + \cdots + b_{n_b} u(t-n_b) + e(t), \tag{3.6}$$

where $y(t)$ is the output at time t, $a_1 \ldots a_n$ and $b_1 \ldots b_n$ are the output and input parameters to be estimated, $y(t-1) \ldots y(t-n_a)$ are the past outputs, $u(t-1) \ldots u(t-n_b)$ are the inputs and $e(t)$ is the white noise disturbance value.

Rearrange the Eq. (3.6), output $y(t)$ in terms of past inputs and past outputs will give:

$$y(t) = a_1 y(t-1) - \cdots - a_{n_a} y(t-n_a) + b_1 u(t-1) + \cdots + b_{n_b} u(t-n_b) + e(t) \tag{3.7}$$

And now, introduce the regression vector:

$$\varphi(t) = \left[-y(t-1) - y(t-2) - \cdots - y(t-n_a) u(t-1) \cdots u(t-n_b)\right]^{\mathrm{T}} \tag{3.8}$$

an adjustable parameter is:

$$\theta = \left[a_1 \ldots .a_{n_\mathrm{a}}, b_1 \ldots .b_{n_\mathrm{b}}\right]^\mathrm{T} \tag{3.9}$$

In matrix form, Eq. (3.7) can be written as Eq. (3.10), known as linear regression:

$$y(t) = \varphi^\mathrm{T}(t)\theta + e(t) \tag{3.10}$$

The output $y(t)$ can be predicted based on regression vector and parameter's vector θ, thus $y(t)$ can be represented as:

$$\hat{y}(t|\theta) = \varphi^\mathrm{T}(t)\theta + e(t), \tag{3.11}$$

where $\varphi(t)$ is called *regression vector,* and its components are the *regressors.* The regress means to calculate the $y(t)$ at a given point in time based on previous values of its inputs and outputs [3].

3.5 Model Validation

The purpose of model validation is to verify that the identified model fulfills the modeling requirements according to objective criteria of good model approximation that can demonstrate the same dynamic characteristics as the real process [3, 12]. To ascertain the quality of the model, the data sequence is divided into two-part, estimation part and validation part or by using one series of experimental data for estimation, another series of data is used for validating purposed [12, 13].

3.5.1 *One-Step-Ahead Prediction*

One-step-ahead prediction (1-SAP) is one of the most common techniques used for model validation. 1-SAP is produced based on the past actual output rather than past prediction output. The expression of 1-SAP is defined as below [14–16]:

$$\begin{aligned} \hat{y} = \hat{f}\Big(&y(t-1), \ldots, y(t-n_y), u(t-1), \ldots, u(t-n_u), \\ &\varepsilon\big(1-1, \tilde{\theta}\big), \ldots, \varepsilon\big(t-n_\varepsilon, \tilde{\theta}\big), \end{aligned} \tag{3.12}$$

where $\hat{f}(\cdot)$ is the estimate of $f(\cdot)$, $\varepsilon\big(t, \hat{\theta}\big)$ is the prediction error or residual.

3.5.2 Residual Analysis

Residual, ε is the difference between the one-step-predicted output from the model and the measured output from the validation data set. Residuals represent the portion of the validation data not explained by the model [3, 11]. Mathematically, residual can be defined as:

$$\varepsilon(t, \theta) = y(t) - \hat{y}(t - \theta), \tag{3.13}$$

where $y(t)$ is the measured output and $\hat{y}$ is the predicted output. The null hypothesis of the residuals is described below and can be used in statistical tests such as autocorrelation test and cross-correlation test [2, 3]:

- ε constitute a white-noise process with mean 0.
- ε are normally distributed.
- ε are symmetrically.
- ε are independent of previous inputs.
- ε are independent of all input if there is no feedback.

3.5.3 Coefficient of Determination

Coefficient of determination also known as model fitting, R^2 is used to measures how successful the fit in explaining the variation within the data and expressed as [11];

$$R^2 = \left[1 - \frac{\sum_{t=n_a}^{n_t} (\hat{\varepsilon}(t))^2}{\sum_{t=n_a}^{n_t} (y(t) - \bar{y})^2} \right] * 100, \tag{3.14}$$

where $\hat{\varepsilon}(t)$ is the error between desired and actual output at time t, $y(t)$ is actual output at time t, $\overline{y}$ is an average for actual output, n_a and n_b are initial and last data, respectively. R^2 takes on values between 0 and 1, usually present in percentage. The higher the R^2 means the more useful the model.

3.5.4 Root-Mean-Square Error

The root-mean-square error (RMSE) is also known as the fit standard error and the standard error of the regression [17]. The RMSE is expressed as [14, 15]:

$$\text{RMSE} = \sqrt{1/N \sum_{i=1}^{N} (y - \tilde{y})^2}. \tag{3.15}$$

RMSE values closer to 0 indicate a fit that is more useful for prediction.

3.5.5 *Correlation Test*

A more convincing method of model validation is to use correlation tests [18]. Two correlation tests are [3, 11]:

- Auto-correlation function (ACF) of the residual for each output and known as the whiteness test.
- Cross-correlation function (CCF) between the input and the residuals and known as the independence test.

The ACF and CCF, respectively, are defined as [3]:

$$R_{\varepsilon\varepsilon}(\tau) = \frac{\sum_{t=1}^{N-\tau}(\varepsilon(t) - \overline{\varepsilon})(\varepsilon(t-\tau) - \overline{\varepsilon})}{\sum_{t=1}^{N}\left(\varepsilon(t - \overline{\varepsilon})^2\right.} \quad (3.16)$$

$$R_{u\varepsilon}(\tau) = \frac{\sum_{t=1}^{N-\tau}\left(u(t - \overline{u})(\varepsilon(t-\tau) - \overline{\varepsilon})\right.}{\sqrt{\sum_{t=1}^{N}(u(t) - \overline{u})^2\sum_{t=1}^{N}(\varepsilon(t) - \overline{\varepsilon})^2}}, \quad (3.17)$$

where $\overline{\varepsilon}$ is the average residuals, $\overline{u}$ is the average input. It is common to investigate between lags $\pm$ 20 [3]. In general, the correlation must be within 95% be confidence intervals, $\pm 1.96/\sqrt{N}$, where N is the total number of data points, then the null hypothesis will be accepted [19]. According to test criteria, a good model has the residual autocorrelation function inside the confidence interval of the corresponding estimates, indicating that the residuals are uncorrelated. According to the independence test criteria, a good model has residuals uncorrelated with past inputs.

3.6 PID Controller

The conventional PID controllers are used in most of all industrial controllers today because of simple structures and ease of implementation. Figure 3.5 illustrates the block diagram of a PID controller modules [20–23];

The PID controller is given in Eq. (3.18) where K_p, K_i, and K_d are the coefficients for proportional gain, integral gain and derivative gain, respectively whereas $u(t)$ is the control signal and $e(t)$ is the error [8]:

$$u(t) = K_p e(t) \int_{t=0}^{\infty} e(t)\mathrm{d}t + K_p K_d \frac{\mathrm{d}e(t)}{\mathrm{d}t} \quad (3.18)$$

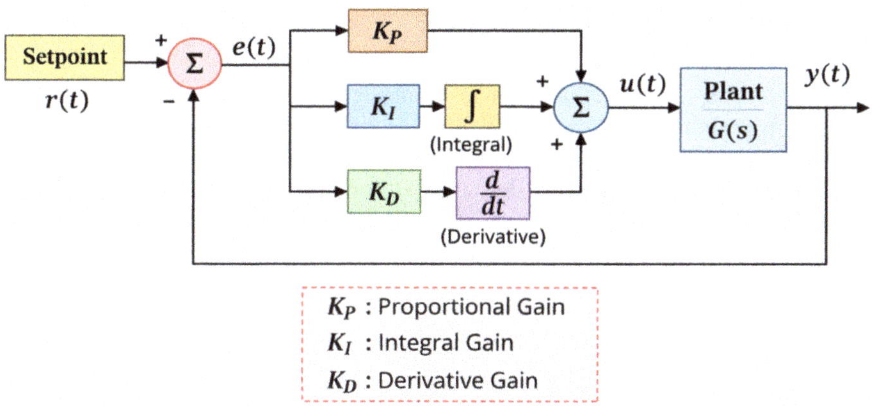

Fig. 3.5 Block diagram of PID controller

Laplace transforms of this equation:

$$\frac{U(s)}{e(s)} = \left(1 + \frac{K_\mathrm{i}}{s} + K_\mathrm{d}s\right)K_p, \tag{3.19}$$

where the constant parameters to the algorithm are expressed as:

$$K_\mathrm{p} = K_\mathrm{c}, K_\mathrm{i} = \frac{1}{T_\mathrm{i}s}, K_\mathrm{d} = T_\mathrm{d}s. \tag{3.20}$$

The functionalities of these three terms are highlighted in Table 3.1 [5]:

The functions of proportional, integral, and derivative controllers complement each other. If they are combined it's possible to make a system that responds quickly to changes (derivative), tracks required position (proportional), and reduces steady-state errors (integral).

Table 3.1 Functionalities of PID Gain

Gain	Functionalities
Proportional	• improving an overshoot and a rise time response
Integral	• eliminates offset • tends to respond slowly at first, but over a long period, they tend to eliminate errors • the integral controller eliminates the steady state error but may make the transient response worse
Derivative	• the controller output is proportional to the rate change during the measurement or error • it improves transient response and the flatness of the response

3.6.1 The Ziegler–Nichols Tuning

Controller tuning means adjustment of controller settings to achieve the desired closed-loop system performance [5]. The Ziegler–Nichols tuning formula is the most well-known tuning method to determine the PID controller parameters depends essentially on the relation between dead time, θ, dominating time constant, τ and static gain, K [8]. This method also known as process reaction curve method which is based on single a single test [5, 24]. In this method, the PID parameters are calculated as [24, 25];

$$\begin{aligned} K_\mathrm{c} &= \tfrac{1.2\tau}{K\theta} \\ T_\mathrm{i} &= 2\theta \\ T_\mathrm{d} &= 0.5\theta, \end{aligned} \tag{3.21}$$

where K_C, $T_\mathrm{i,}$ and T_d denote as controller gain, integral time, and derivative time, respectively.

3.7 Basic Concepts of the Fuzzy Set

The fuzzy sets are a mathematical means to represent vague information and a collection of real numbers having the partial degree of membership [2]. Two factors need to be considering in specify a fuzzy set: (1) the members from the set and (2) each member's degree of membership in the set. The membership to the set is specified by a membership value between 0 and 1.

Fuzzy sets and fuzzy logic are used to heuristically quantify the meaning of linguistic variables and linguistic values [2, 26]. If the physical quantity under consideration is described by a word such as temperature, rather than symbol (say, t) that world is called a linguistic variable [2]. The linguistic variables are needed to describe fuzzy system inputs and outputs, and the characteristics of the inputs and outputs. Fuzzy sets are also usually given linguistic names, called linguistic values. For instance, for the "temperature" variable, a fuzzy set can be defined with linguistic values "cold," "cool," "warm," and "hot." Linguistic names are used for variables and their values in fuzzy logic because people usually think and speak in linguistic terms [2].

In practical applications, the universe of discourse means the collection of numbers on which a variable is defined, usually some interval of real numbers. For example, the effective universe of discourse for temperature would be in the interval [0, 100] °C. The effective universe of discourse $[\alpha, \beta]$ means where α and β are the points at which the outermost membership functions saturate for input's universe of discourse, or the point beyond which the outputs will not move through the output universe of discourse [2].

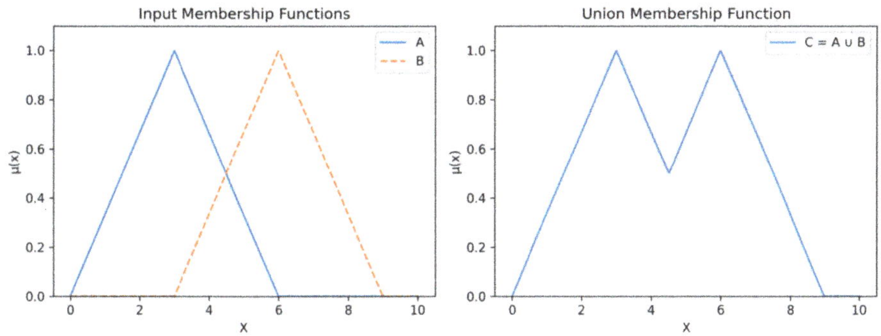

Fig. 3.6 Union membership function

3.7.1 *Fuzzy Union (or)*

Let M^A and M^B be two fuzzy sets defined as a variable on the universe of discourse of χ, and let their associated membership functions be μ^A(x) and μ^B(x), respectively. The union of fuzzy sets M^A and M^B denoted by $M^A \cup M^B$. Figure 3.6 depicts the membership function in the union of two membership functions [2]. The membership function can be defined by one of the following methods [2];

1. *Maximum*: Find the maximum of the membership values as in.

$$\mu^{M^A \cup M^B}(x) = \max\{\mu^A(x), \mu^B(x) : x \in \chi\}. \tag{3.22}$$

2. *Algebraic sum*: Find the algebraic sum of the membership values as in.

$$\mu^{M^A \cup M^B}(x) = \{\mu^A(x) + \mu^2(x) - \mu^B(x)\mu^B(x) : x \in \chi\}. \tag{3.23}$$

3.7.2 *Fuzzy Intersection (AND)*

Let M^A and M^B be two fuzzy sets defined as a variable on the universe of discourse of χ, and let their associated membership functions be μ^A(x) and μ^B(x), respectively. The fuzzy intersection of M^A and M^B denoted by $M^A \cap M^B$. Figure 3.7 shows the intersection between two membership functions [2]. The membership function can be defined by one of the following methods [2];

1. *Minimum*: Find the minimum of the membership values as in.

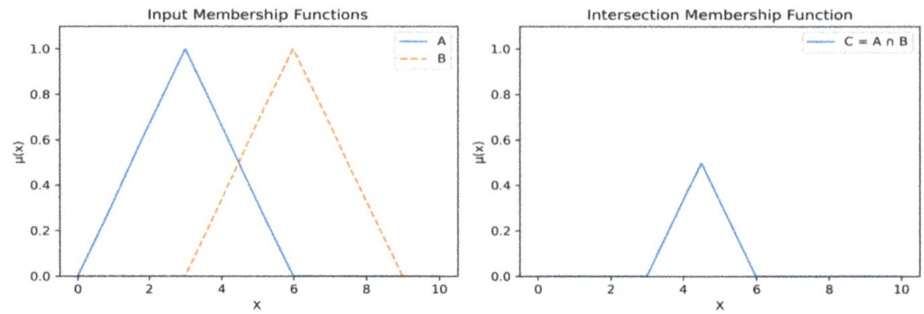

Fig. 3.7 Intersection membership function

$$\mu^{M^A \cap M^B}(x) = \min\{\mu^A(x), \mu^B(x) :\in \chi\}. \tag{3.24}$$

2. Algebraic Product: Find the product of the membership values as in.

$$\mu^{M^A \cap M^B}(x) = \{\mu^A(x)\mu^B(x) : x \in \chi\} \tag{3.25}$$

3.8 Fuzzy Logic Controller

Fuzzy logic controllers are more robust than PID controllers because this controller can cover a much wider range of operating conditions and can operate with white-noise disturbances of different natures [27]. The architecture of FLC is given in Fig. 3.8 where the plant output is denoted by $y(t)$, input to the plant is $u(t)$ and the reference input to the FLC is denoted by $r(t)$ [2].

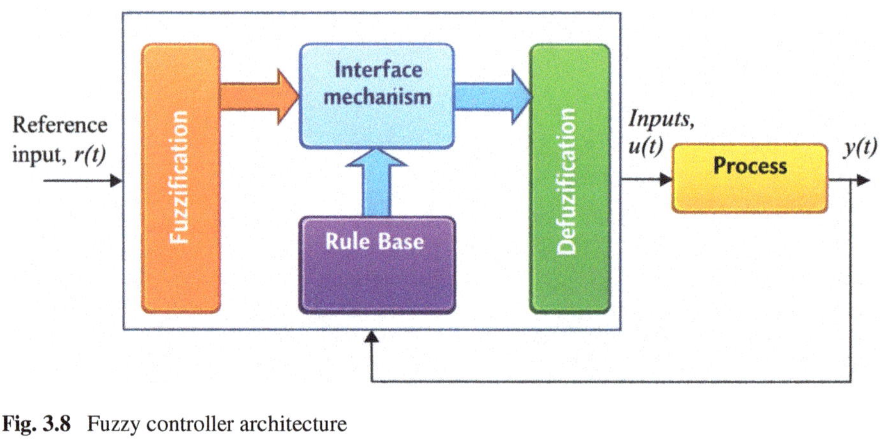

Fig. 3.8 Fuzzy controller architecture

The inputs, $u(t)$ and reference input, $r(t)$ are crisp values, which are real numbers, not fuzzy sets [2, 27]. The FLC consists of fuzzification, rule-base, inference mechanism and defuzzification components. The fuzzification converts controller inputs into information that the inference mechanism can easily use to activate and compared with the rules in the rule-base. A rule-base holds the knowledge of the expert's linguistic description of how best to control the system. An inference mechanism also called inference engine, or fuzzy inference evaluates which control rules are relevant to the current time and then decide what the input to the plant should be. The defuzzification converts the conclusions about the inference mechanism into crisp real value for a fuzzy output [2, 28–30].

3.8.1 Selection of Controller Input

In designing the fuzzy logic controller, the first step is to take the inputs and determine the degree to which they belong to each of the appropriate fuzzy sets via membership functions. The input is always a crisp numerical value limited to the universe of discourse of the input variable [27, 28]. However, the choice of the inputs to the controller is a crucial job. This is because to make sure the controller will have the proper information to make good decisions and have proper control inputs to be able to maintain the system at the desired and achieve high-performance operation. Practically, for most control applications, two inputs were employed, namely error signal, $e(t)$ and change of error $\Delta e(t)$. These inputs can describe the condition of the step response, whether the error is negative or positive [9].

3.8.2 Membership Function

A membership function (MF) is a curve that defines how each point in the input space is mapped to a membership value or degree of membership. The input space is referred to as the universe of discourse. The MF must vary between 0 and 1. The membership function is often given the designation of μ [27].

There are many other's choices for the shape of membership function are possible such as:

- Piecewise linear (triangular and trapezoidal);
- Gaussian distribution function;
- Quadratic and cubic polynomial curves;
- Sigmoid curve.

Different shapes will provide different meaning for the linguistic values that they quantify [27]. Different researchers choose numerous shapes in various application problems. In most applications of identification and control, triangular and

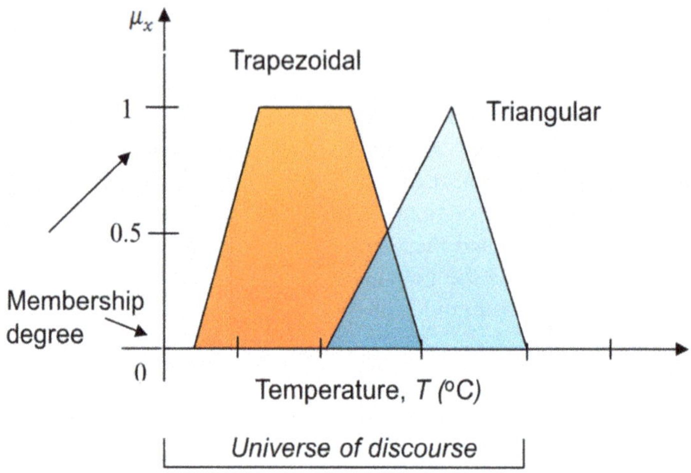

Fig. 3.9 Triangular and trapezoidal membership function shapes

trapezoidal membership functions have proven to be more popular with fuzzy logic theoretic and practitioners. Their reasons are [27]:

- Simplicity of this function often allows for the prediction and calculation of an output of the fuzzy system.
- The extra smoothness introduced by higher-order fuzzy sets and demanding higher computational consumption is not strongly reflected in the quality of a fuzzy model.

Figure 3.9 shows the triangular and trapezoidal membership function shapes [2]: In the membership function choice, a user can be advised as follows [27, 28];

- Initially, one chooses the width of the membership functions to provide the whole overlap about 12–14%.
- To improve steady-state accuracy one must decrease the membership functions' whole overlap after their initial choice and simulation.
- To improve stability characteristics such as oscillation, settling time and overshoot, one must increase the whole overlap.
- The use of a fuzzy controller with wider membership functions and higher overlap can be recommended in the presence of high disturbances.

3.8.3 Interface Rules

The rule-based is a set of linguistic statement. In rule-based inference systems, the universe is partitioned using concepts, modeled via sets. Reasoning is then based on expressions of logical relationships between the concepts: "IF-THEN" rules. Most if-then rules used in fuzzy control and identification, and represented such as in Eq.

(3.26) [28–30];

$$\text{if premise then consequent} \tag{3.26}$$

$$R_i \text{ If } \tilde{x} \text{ is } \tilde{P}, \text{ then } \tilde{y} \text{ is } \tilde{Q}, \tag{3.27}$$

where $\tilde{x}$ is a linguistic variable defined on universe χ, $\tilde{P}$ is a linguistic value described by fuzzy set P defined on universe χ, $\tilde{y}$ is a linguistic variable on universe γ, and $\tilde{Q}$ is a linguistic value described by fuzzy Q defined on universe γ. The first part of the statement $\tilde{x}$ is $\tilde{P}$ is called the premise of the rule, sometimes called antecedents and the second part of the statement $\tilde{y}$ is $\tilde{Q}$ is called the consequent of the rule, sometimes called actions. In a fuzzy system, sometimes there may be more than one part upon the premise, which is the rule R_j denoted as [2, 28];

$$R_j \text{ If } \tilde{x}_1 \text{ is } \tilde{P}_1 \text{ and } \tilde{x}_2 \text{ is } \tilde{P}_2 \text{ then } \tilde{y} \text{ is } \tilde{Q}. \tag{3.28}$$

For example: **If** error is poslarge **and** change in error is negsmall **then** temperature is negsmall.

The advantages of if-then rule structure are convenient in that it is easily comprehensible as being close to one of the ways humans store knowledge, it also provides explanations for the model outcome since it is always possible to find out the exact rules that were fired and to convert these into semantically meaningful sentences.

3.8.4 Defuzzification

The function of the defuzzification is to convert the collection of recommendations of all rules into crisp real value for a fuzzy output. To get one crisp output, all the recommendations will combine by taking a weighted average of the various recommendations. The choice of the defuzzification procedure depends mainly on personal preference. There are many methods in defuzzification procedure such as mean of maxima, height defuzzification, center average (CA), first-of-maxima, and center-of-largest-area defuzzification. However, the center-of-area method also known as the center-of-gravity (COG) method is the most well-known defuzzification method [2, 28]. This method gives better accuracy and fast computation time [9]. The COG method determines the center of the area below the combined membership function. Equation (3.29) expressed the COG defuzzification as [9]

$$U = \frac{\sum_{i=1}^{l} U_i \mu(y)}{\sum_{i=1}^{l} \mu(y)}, \tag{3.29}$$

where U_i is the center of the output membership function, and $\mu(y)$ is the degree of fulfillment.

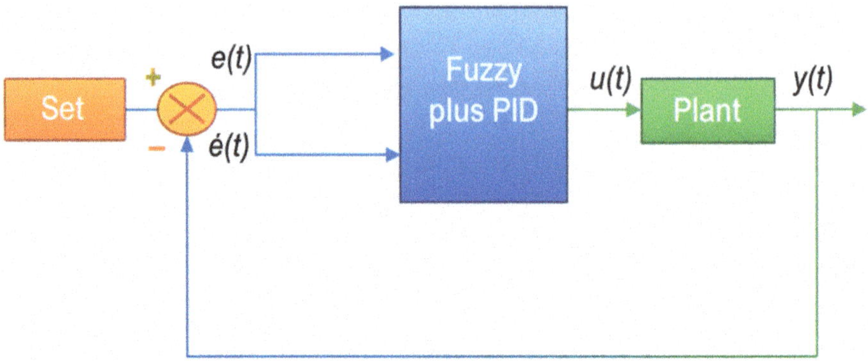

Fig. 3.10 Block diagram of hybrid fuzzy plus PID

3.9 Hybrid Fuzzy Plus PID Controller

Hybridization of FLC and PID controller is the process where two controllers are combined together to improve system performance at a transient and steady state by exploit beneficial sides of both controllers [31]. Figure 3.10 shows the general block diagram of HFPID controller [32]:

There are eight gain parameters need to be tuned properly in the HFPID which are K_{p1}, K_{d1}, G_{f1}, G_{f2}, G_{f3} is fuzzy gains and G_e, K_{i2}; K_{d2} is PID gains. This HFPID controller using two inputs, which are the $e(t)$ and change of error, $\dot{e}(t)$ and one output, $u(t)$

3.10 Self-tuning Fuzzy PID Controller

The self-tuning fuzzy PID controller is the controller that applied the fuzzy tuner to tune automatically the PID coefficients, K_P, K_I, and K_D [33–35]. The block diagram of STFPID controller is depicted in Fig. 3.11 [34]. There are two inputs to the fuzzy controller which are errors, $e(t)$ and derivative error, Δe, and three output for each respective PID coefficients', $K_I^{'}$, and $K_D^{'}$. Error, $e(t)$ is calculated upon the difference between set point a set point actual output [36, 37]. The interval for parameters K_p, $K_{i,}$ and K_d supposedly in $[K_{i\,min}, K_{i\,max}]$ and $[K_{d\,min}, K_{d\,max}]$ forms [34, 38]. The PID parameters automatically updates as the response change based on the $e(t)$ and Δe [34, 38].

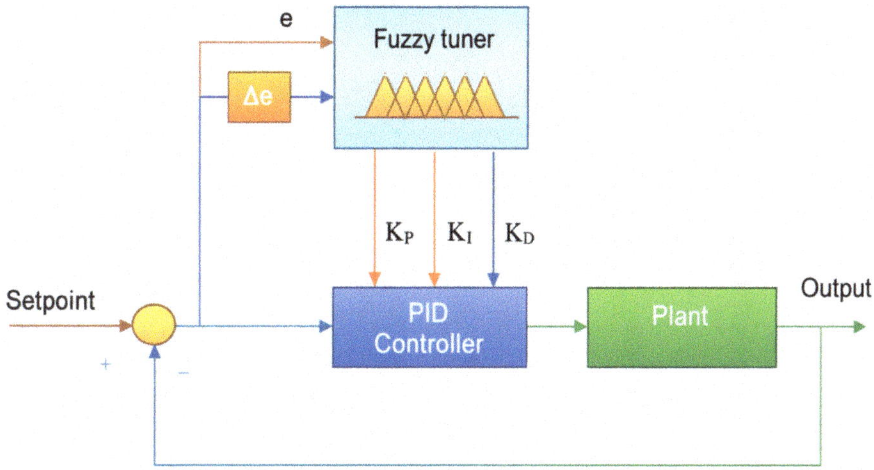

Fig. 3.11 The block diagram of self-tuning fuzzy PID controller

3.11 Performance Evaluation

Ideally, the aims of control system applications and design the controllers is to ensure the close loop system satisfy performance criteria such as the system must be stable, minimized the effects of disturbances, good set-point tracking, which is rapid and smooth response to set point changes [5]. To optimize the closed-loop response, performance evaluation was used by measuring the step response to the controller. The standard performance criteria used in the system's step response to represent dynamic performance are percentage overshoot, rise time, and settling time. Figure 3.12 shows the step response performance evaluation [9];

3.11.1 Percentage Overshoots

Percentage overshoots, %OS is the maximum peak value on the response curve measured from unity (final value) [2, 14, 39];

$$\%OS = \exp\left(\frac{-\zeta\pi}{\sqrt{1-\zeta^2}}\right) \times 100, \tag{3.30}$$

where ζ is a damping ratio.

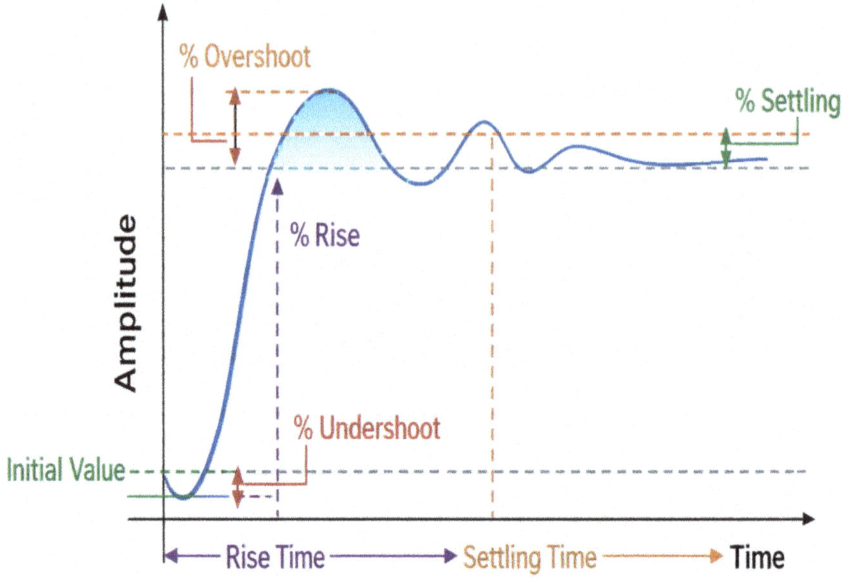

Fig. 3.12 Evaluation of step response performance

3.11.2 Rise Time

Rise time, t_r is the time required for the response to have the value from 10 to 90% or 5–95% or 0–100% of its final value. It can be divided into two depending upon the type to the system as follows [2]:

- t_r for under damped normally 0–100%,
- t_r for over damped normally 10–90%

$$t_r = \frac{\pi - \beta}{\omega_d}, \tag{3.31}$$

where

$$\beta = \tan^{-1}\left(\frac{\sqrt{1-\zeta^2}}{\zeta}\right) \tag{3.32}$$

$$\omega_d = \omega_n\sqrt{1-\zeta^2}, \tag{3.33}$$

where ω_n is the natural frequency.

3.11.3 Settling Time

Settling time, t_s is the time take by response curve to reach and stay within a range of the final value [2];

$$t_s = \frac{4}{\zeta \omega_n} \text{for } (2\%) \tag{3.34}$$

$$t_s = \frac{3}{\zeta \omega_n} \text{for } (5\%), \tag{3.35}$$

Generally, this chapter explains the theoretical background about the linear system identification and controller design. The input signal used in system identification can have a significant influence on the estimated model, and some aspects need to be considered in selecting of input signal. In modeling section, there are several steps need to be followed such as model structure selection, model estimation and model validation. In model validation, some technique such as 1-SAP, residual analysis, determination of coefficient, RMSE, ACF, and CCF were used to verify that the identified model fulfills the modeling requirements. In the following section, the structure and working principle of PID controller were discussed. The Ziegler–Nichols tuning formula is the most well-known tuning method to determine the PID controller parameters depends essentially on the relation between dead time, dominating time constant, and static gain. Finally, this chapter closed with the basic concept of a fuzzy logic controller consists of fuzzification, rule-base, inference mechanism and defuzzification components.

References

1. Balakrishnan, V.: System identification: theory for the user (second edition). Automatica. **38**(2), 214–219 (2002). https://doi.org/10.1016/s0005-1098(01)00214-x
2. Yusoff, Z.M.: Self-tuning Fuzzy PID Control of Hydro-diffusion Essential Oil Extraction System. Universiti Teknologi MARA, Malaysia (2014)
3. Perspectives on system identification: LLjung, L. Annu. Rev. Control. **34**, 1–12 (2010). https://doi.org/10.1016/j.arcontrol.2009.12.001
4. Ljung, L., Vicino, A.: Guest editorial: special issue on system identification. IEEE Trans. Autom. Control **50**(10), 1473–(2005). https://doi.org/10.1109/TAC.2005.856638
5. Bala, B.K.: Process dynamics and control. In: Agro-product Processing Technology. CRC Press, Boca Raton (2020). https://doi.org/10.1201/9780429487507-14
6. Maiti, S.N., Saraf, D.N.: Adaptive dynamic matrix control of a distillation column with closed-loop online identification. J. Process Control. **5**(5), 315–327 (1995). https://doi.org/10.1016/0959-1524(95)00002-8
7. Ishak, A.A., Ahmad, A.: Application of numerical technique in tangent analysis of a discrete step response data. In: 2002 Student Conference on Research and Development: Globalizing Research and Development in Electrical and Electronics Engineering (SCOReD 2002)—Proceedings (2002). https://doi.org/10.1109/SCORED.2002.1033045

8. Hägglund, T.: An industrial dead-time compensating PI controller. Control Eng. Pract. **4**(6), 749–756 (1996). https://doi.org/10.1016/0967-0661(96)00065-2
9. Yusuf, Z., Janin, Z., Taib, M.N.: Self Tuning Fuzzy Logic controller with Smith structure for controlling glycerin bleaching temperature. In: Proceedings of the ICSGRC 2010: 2010 IEEE Control and System Graduate Research Colloquium (2010). https://doi.org/10.1109/ICSGRC.2010.5562526
10. Norton, J.P.: System identification and control design using P.I.M. + software. I. D. Landau. Automatica. **27**(6), 1056–1057 (1991). https://doi.org/10.1016/0005-1098(91)90148-U
11. Ljung, L.: System identification toolbox. In: The Matlab User's Guide, vol. 1 (2011)
12. Sohlberg, B.: Grey box modelling for model predictive control of a heating process. J. Process Control. **13**(3), 225–238 (2003). https://doi.org/10.1016/S0959-1524(02)00030-6
13. Shojaeefar, M.H., Goudarzi, K., Noorpoor, A.R., Fazelpour, M.: A study of thermal contact using nonlinear system identification models. Am. J. Eng. Applied Sci. **1**(1), 16–23 (2008). https://doi.org/10.3844/ajeassp.2008.16.23
14. Yusoff, Z.M., Muhammad, Z., Rahiman, M.H.F., Taib, M.N.: ARX modeling for down-flowing steam distillation system. In: Proceedings of the 2012 IEEE 8th International Colloquium on Signal Processing and Its Applications (CSPA 2012) (2012). https://doi.org/10.1109/CSPA.2012.6194774
15. Yusoff, Z.M., Muhammad, Z., Rahiman, M.H.F., Taib, M.N.: PRBS perturbation and modeling of hydro-diffusion steam distillation essential oil extraction system under feed-in disturbance. In: Proceedings of the 2012 IEEE 8th International Colloquium on Signal Processing and Its Applications (CSPA 2012) (2012). https://doi.org/10.1109/CSPA.2012.6194773
16. Ismail, N., Rahiman, M.H.F., Taib, M.N.: Noise variation test for ARX model of a heating process in essential oil extraction plant. In: Proceedings of the CSPA 2010: 2010 6th International Colloquium on Signal Processing and Its Applications (2010). https://doi.org/10.1109/CSPA.2010.5545304
17. MathWorks: Curve Fitting Toolbox^{TM}: User's Guide (2011)
18. Nooraii, A., Romagnoli, J.A., Figueroa, J.: Process identification, uncertainty characterisation and robustness analysis of a pilot-scale distillation column. J. Process Control 9(3), 247–264. https://doi.org/10.1016/S0959-1524(98)00042-0
19. Ahmad, S.M., Shaheed, M.H., Chipperfield, A.J., Tokhi, M.O.: Nonlinear modelling of a twin rotor MIMO system using radial basis function networks. In: National Aerospace and Electronics Conference, Proceedings of the IEEE (2000). https://doi.org/10.1109/naecon.2000.894926
20. Yusoff, Z.M., Muhammad, Z., Rahiman, M.H.F., Taib, M.N.: Real time steam temperature regulation using self-tuning fuzzy pid controller on hydro-diffusion essential oil extraction system. Int. J. Smart Sensing Intell. Syst. **6**(5), 2055–2074 (2013). https://doi.org/10.21307/ijssis-2017-626
21. Tajjudin, N., Yusuf, Z., Rahiman, M.H.F., Taib, M.N.: Self tuning PID controller for steam distillation essential oil extraction system. In: Proceedings of the 6th International Colloquium on Signal Processing and Its Applications (CSPA 2010) (2010). https://doi.org/10.1109/CSPA.2010.5545292
22. Muhammad, Z., Jamil, K.A., Yusoff, Z.M., Rahiman, M.H.F.: Application of self-tuning fuzzy PID (STFPID) controller on industrial essential oil extraction system using system identification approach. J. Telecommun. Electron. Comput. Eng. **10**(1–3), 77–81 (2018)
23. Yusoff, Z.M., Muhammad, Z., Nordin, M.N.N., Rahiman, M.H.F., Taib, M.N.: Real time PID control for hydro-diffusion steam distillation essential oil extraction system using gradient descent tuning method. In: Proceedings of the 2012 IEEE Control and System Graduate Research Colloquium, ICSGRC 2012 (2012). https://doi.org/10.1109/ICSGRC.2012.6287178
24. Tavakoli, S., Tavakoli, M.: Optimal tuning of PID controllers for first order plus time delay models using dimensional analysis. In: International Conference on Control and Automation (2003). https://doi.org/10.1109/icca.2003.1595161
25. Tajjudin, M., Ishak, N., Ismail, H., Rahiman, M.H.F., Adnan, R.: Optimized PID control using Nelder-Mead method for electro-hydraulic actuator systems. In: Proceedings of the 2011 IEEE

Control and System Graduate Research Colloquium (ICSGRC 2011) (2011). https://doi.org/10.1109/ICSGRC.2011.5991836
26. Chen, W.C., Chang, N.B., Chen, J.C.: Rough set-based hybrid fuzzy-neural controller design for industrial wastewater treatment. Water Res. **37**(1), 95–107 (2003). https://doi.org/10.1016/S0043-1354(02)00255-5
27. Jang, J.S.R., Gulley, N.: Fuzzy Logic Toolbox User's Guide, vol. 1, no. 995. The Mathworks Inc. (2015)
28. Lee, C.C.: Fuzzy logic in control systems: fuzzy logic controller—Part I. IEEE Trans. Syst. Man Cybern. **20**(2), 404–418 (1990). https://doi.org/10.1109/21.52551
29. Visioli, A.: Tuning of PID controllers with fuzzy logic. IEE Proc. Control Theory Appl. **148**(1), 1–8 (2001). https://doi.org/10.1049/ip-cta:20010232
30. Eker, I., Torun, Y.: Fuzzy logic control to be conventional method. Energy Convers. Manag. **47**(4), 377–394 (2006). https://doi.org/10.1016/j.enconman.2005.05.008
31. Erenoglu, I., Eksin, E., Yesil, E., Guzelkaya, M.: An intelligent hybrid fuzzy PID controller. In: 20th European Conference on Modelling and Simulation: Modelling Methodologies and Simulation Key Technologies in Academia and Industry (ECMS 2006) (2006). https://doi.org/10.7148/2006-0062
32. Kasuan, N., Yusuf, Z., Taib, M.N., Rahiman, M.H.F., Tajuddin, N., Aziz, M.A.A.: Robust steam temperature regulation for distillation of essential oil extraction process using hybrid fuzzy-PD plus PID controller. World Acad. Sci. Eng. Technol. **47**, 102–107 (2010)
33. Ahn, K.K., Truong, D.Q.: Online tuning fuzzy PID controller using robust extended Kalman filter. J. Process Control. **19**(6), 1011–1023 (2009). https://doi.org/10.1016/j.jprocont.2009.01.005
34. Rahmat, M.F., Has, Z.: Application of self-tuning fuzzy PID controller on industrial hydraulic actuator using system identification approach. Int. J. Smart Sensing Intell. Syst. **2**(2), 250–270 (2009). https://doi.org/10.21307/ijssis-2017-349
35. Ahn, K.K., Nguyen, B.K.: Position control of shape memory alloy actuators using self tuning fuzzy PID controller. Int. J. Control Autom. Syst. **4**(6), 704–713 (2006)
36. Gao, Z., Trautzsch, A.A., Dawson, J.G.: A stable self-tuning fuzzy logic control system for industrial temperature regulation. IEEE Trans. Indus. Appl. **38**(2), 414–424 (2002). https://doi.org/10.1109/28.993163
37. Carvajal, J., Chen, G., Ogmen, H.: Fuzzy PID controller: design, performance evaluation, and stability analysis. Inf. Sci. (N. Y.). **123**(3–4), 249–278 (2000). https://doi.org/10.1016/S0020-0255(99)00127-9
38. Sinthipsomboon, K., Hunsacharoonroj, I., Khedari, J., Pongaen, W., Pratumsuwan, P.: A hybrid of fuzzy and fuzzy self-tuning PID controller for servo electro-hydraulic system. In: Proceedings of the 2011 6th IEEE Conference on Industrial Electronics and Applications, ICIEA 2011 (2011). https://doi.org/10.1109/ICIEA.2011.5975583
39. Muhammad, Z., Yusoff, Z.M., Rahiman, M.H.F., Taib, M.N.: Modeling of steam distillation pot with ARX model. In: Proceedings of the IEEE 8th International Colloquium on Signal Processing and Its Applications (CSPA 2012) (2012). https://doi.org/10.1109/CSPA.2012.6194717

Chapter 4
From Concept to Experiment: Designing and Implementing the Intelligent Extraction Control System

Abstract This chapter presents the experimental setup, data acquisition procedure, and system modeling of a temperature-controlled essential oil extraction process. A laboratory-scale distillation system is developed to facilitate controlled experimentation, incorporating key components such as a heating element, temperature sensors, and a data acquisition system for real-time monitoring. The system is designed to enable accurate measurement and recording of process variables required for modeling and control purposes. To capture the dynamic behavior of the extraction system, appropriate excitation signals are applied to the heating process, and input–output data are collected under controlled operating conditions. The acquired data are then pre-processed and used for system identification, where mathematical models are developed to represent the relationship between input power and output temperature. Both linear and nonlinear modeling approaches are considered to accurately describe the system dynamics. Model validation is performed using statistical and performance-based evaluation methods, including prediction accuracy, error analysis, and correlation tests, to ensure the reliability of the developed models. The resulting models provide a basis for subsequent controller design and performance evaluation.

4.1 Introduction

This chapter is basically divided into four main sub-chapters, which are prototype development, data collection and data analysis, linear system identification, and controller design. In the prototype development section, the detail fabrication of the plant will be explained such as steam distillation essential oil extraction system, temperature instrumentations and voltage-temperature calibration curve. Data collection and data analysis will be covered in the step test, the PRBS characteristics and data pre-processing. Data organization, model structure selection, model selection and model validation will be discussed in linear system identification sub-section. The controller design topic explains deeply on simulation and real time implementation. Finally is the procedure for essential oil extraction.

Z. Mohd Yusoff et al., *Fuzzy Logic, Smart Control, and the Future of Essential Oils*, Studies in Fuzziness and Soft Computing 442,
https://doi.org/10.1007/978-981-92-2084-7_4

4.2 Prototype Development

There are two categories involved throughout the prototype development which are steam distillations essential oil extraction system set up and temperature instrumentations as shown in Fig. 4.1 [1]. In this work, the system is called hydro-diffusion steam distillation essential oil extraction system.

4.2.1 The Hydro-diffusion Steam Distillation Essential Oil Extraction System

The steam distillation is a widespread method to isolate the essential oil from aromatic plants. Although the conventional steam distillation system is characterized by a simple process, cleaner production and low operational costs but unfortunately, losses of some volatile compounds, toxic solvent residue and volatile compounds were significantly diluted with water and will be forever lost is undeniable [2, 3]. The hydro-diffusion steam distillation essential oil extraction system developed in this work as a viable alternative to overcome these shortcomings. The present hydro-diffusion is a small-scale tray, simple, easily constructed apparatus was inspired by the existing steam distillation system, but with different construction. This plant was built with the aims to isolate the essential oil by steam distillation technique by yield

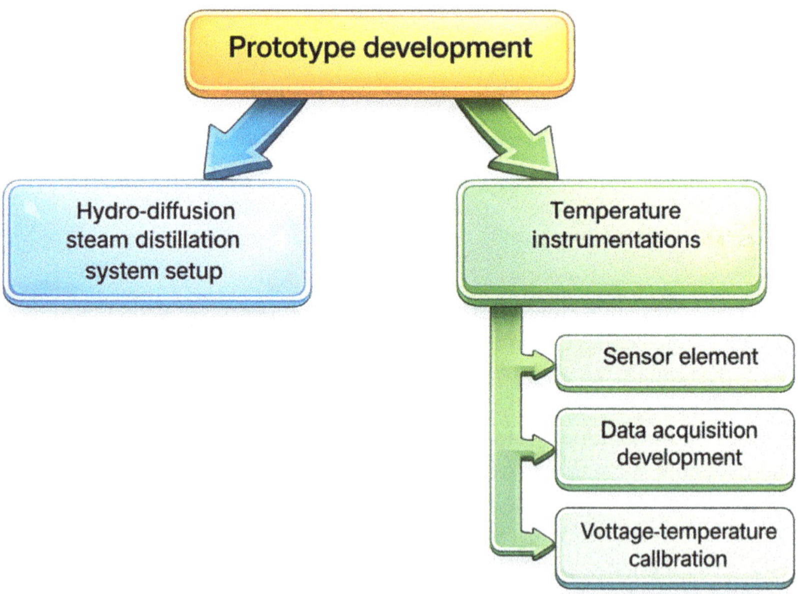

Fig. 4.1 Procedure of prototype development

the best quality of extracts and get high quantity of essential oil as compared with the conventional steam distillation system.

This work demonstrates a new method for extracting the essential oils. The study applied concept of hydro diffusion where steam and extract (water and essential oil) move thus naturally downwards by earth gravity to a condenser [4]. It's opposite to the conventional steam extraction where steam which acts as a transportation medium carry oil goes upward. By using conventional steam distillation, the steam is not fully used for extraction. As steam is passed through the material tray, it condenses in the first element before going to the next and so on, where it releases its enthalpy of vaporization. The steams that release their enthalpy have the probability to go back to the distillation column. This can be proven by examining the water in the distillation column. With long extraction time the water becomes darker. The hydro-diffusion steam distillation plant enhances oil recovery by minimizing oil waste in boiling water. Some literature mentioned that by using a conventional steam distillation process, residual oil dissolved in the water will cause an odor nuisance and waste of the valuable product in the water stream [5]. Redistilling or cohobating is the process to optimize the recovery of essential oil in the wastewater. However, the process offers high cost, mainly heating and energy costs [6]. The hydro-diffusion technique can prevent pollution and at the same time maintain the quality of essential oil [7].

4.2.2 Temperature Instrumentations

This section covers the setup of the sensor used, software development and voltage-temperature calibration.

4.2.2.1 Sensor Device

The sensor element used in this study is RTD PT-100 and SENECA K109T as a signal conditioning to convert the signal from voltage to temperature or vice versa.

RTD Pt-100

The steam temperature and water temperature were measured using RTD PT-100 sensors. RTD PT-100 means a resistance temperature detector that exhibit 100 Ω at 0 °C and 138.4 Ω at 100 °C by using platinum as a metal. The bridge output is in resistive value and was calculated by change in resistance is a function of temperature. The linear approximation of the resistance versus temperature is shown in Eq. 4.1 [1]:

$$R(T) = R(T_o)[1 + \alpha_o \Delta T] \tag{4.1}$$

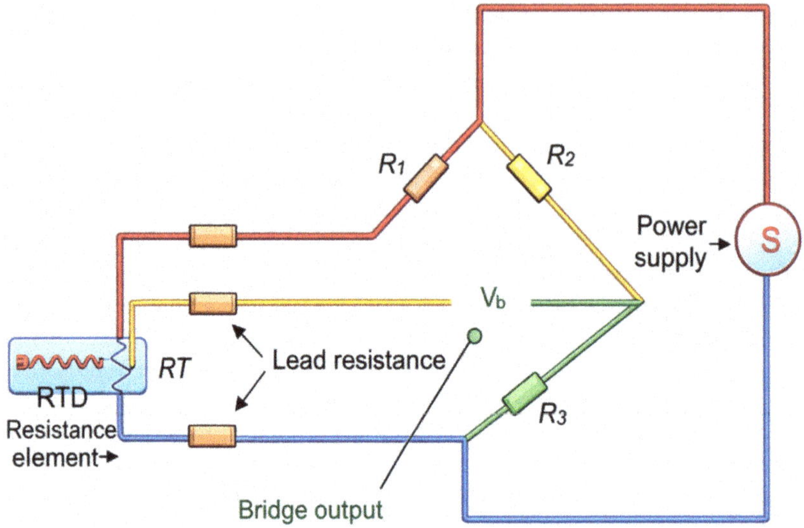

Fig. 4.2 RTD-PT-100 for three wire configurations

where

$R(T)$	approximation of resistance at temperature T
$R(T_0)$	resistance at temperature T_o
Δt	$T - T_o$
α_0	fractional change in resistance per degree of temperature at T_o.

The value of $\alpha_0 = 0.0392$ is taken from resistance versus temperature table. This sensor is available with two, three and four-wire leads. In this work, a three-wire configuration was used where it provides one connection to one end of the element and two to the other end of the element. The connection of the RTD PT-100 for three-wire bridge configuration is demonstrated in Fig. 4.2 as below [1]:

Signal Conditioning SENECA K109T

The SENECA K109PT is a converter that converts a temperature signal read by RTD PT-100 with a connection of 3 wires into a current or voltage standard output signal as shown in Fig. 4.3 [1]. This module with 102.5 × 93.1 × 6.2 mm was designed with the features of compact size, super slim enclosure, fast response time, bus-connector power supply option and completely configurable through two DIP switches. The K109PT with three-way galvanic isolation between input, output and power supply assures the integrity of data collected and can be placed at a considerable distance up to a few hundred meters from the probe.

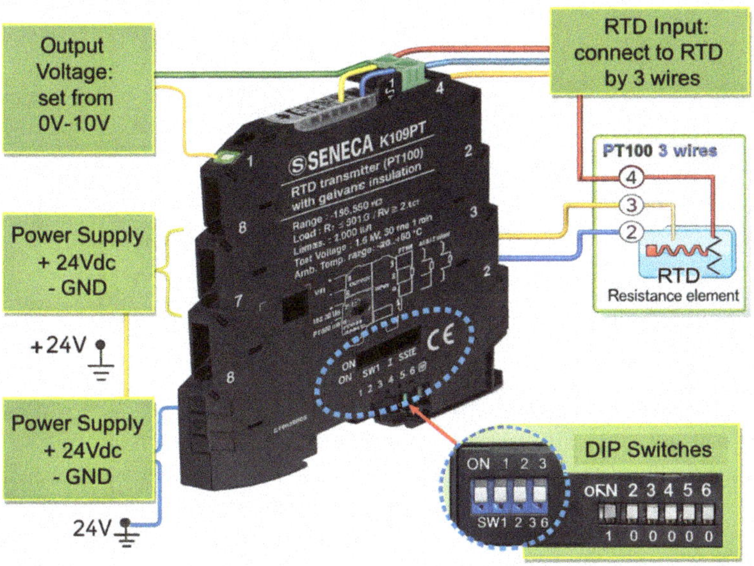

Fig. 4.3 SENECA K109T

4.2.2.2 Data-Acquisition Development

A computer and a data acquisition (DAQ) were used to log in the input-output data from hydro-diffusion steam distillation system. The DAQ programming was developed using MATLAB version R2009a before it was attached to the Advantech PCI-171 interface board. For the calibration purposed, the PCI-1711 is automatically calibrated before logging the input-output data, which has been provided by the hardware. The data acquisition was performed using analogue input (AI) and analogue output (AO) ports, which have been equipped in the PCI-1711. The input-output data were saved in a mat file format with the 1 s sampling period and open loop system.

4.2.2.3 Voltage-Temperature Calibration

The objective of the voltage-temperature calibration is to produce a mathematical equation relating a variable to be measured i.e., temperature, to the electrical signal which is the voltage from 0 to 5 V.

- Input data = temperature, T (°C).
- Output data = voltage (V).

The experiment setup for voltage-temperature calibration curve was done as shown in Fig. 4.4 [1]:

The procedure for voltage-temperature calibration as below:

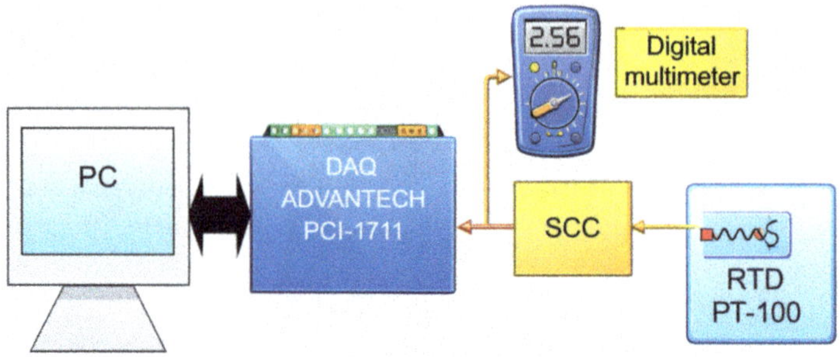

Fig. 4.4 Experiment setup for voltage-temperature calibration curve

1. The calibration Simulink block was developed using the MATLAB programming that will automatically measure the temperature corresponding to the output voltage as illustrated in Fig. 4.5 [1]. From the Fig. 4.5, the gain block that converts the voltage to temperature is set 20 based on the data sheet of SENECA K109 PT. This is because the minimum temperature that can be measured is 0 °C to maximum temperature 200 °C.
2. The experiment starts by placed the RTD PT-100 in the hot water.
3. The output voltage from RTD PT-100 will monitor by a digital multimeter. The data will be measured for every 5 °C, starting from temperature 50 °C.
4. To get the actual temperature, the voltage measured by RTD PT-100 will be recorded and compared with the RTD PT-100 conversion table.
5. As the input and output data were obtained, polynomial linear curve fitting (PLCF) was performed by using curve fitting technique. The temperature relationship as the following form:

$$T = aV^2 + bV + c \tag{4.2}$$

where *a, b,* and *c* are the coefficients for temperature measurement.

Flowchart for PLCF method [1]

See Fig. 4.6.

4.3 Data Collection and Analysis

The data collection is a challenging part because the collected data must be informative enough to represent the process dynamics of the system interest. The experiments should be repeated in the case of insufficient data and non-informative. The data should be pre-processed before further used especially for system identification purposed to obtain the most valuable information from the measured data. However,

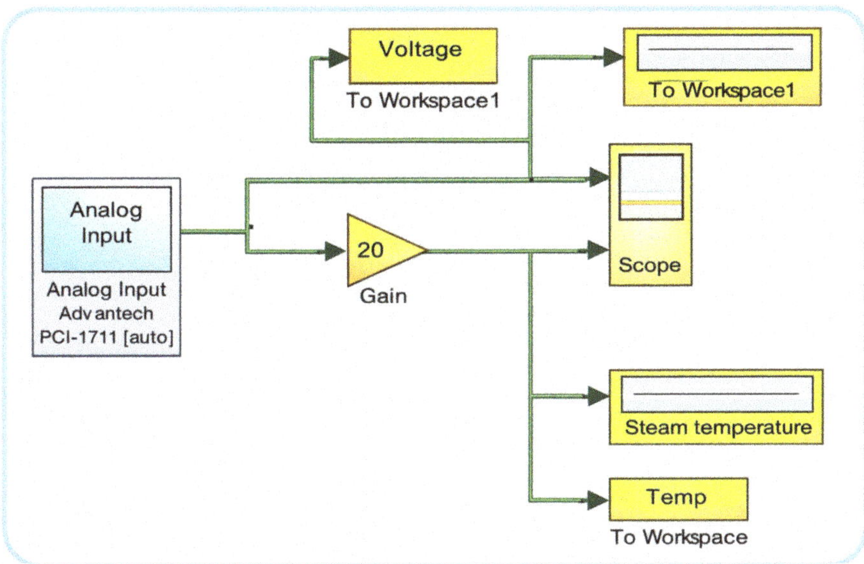

Fig. 4.5 Calibration program

the types of pre-processed data are depending on the system itself. In this work, the data pre-processing involves such as unit conversion and data separation. The prior knowledge is very valuable in system identification stages since it will give some clues in performing the model.

The data collection is divided into two main parts, which are step test then followed by PRBS test. Step test is the conventional method which conducted to observe the process dynamics of heating and cooling process. Step test analysis provides two the most vital information's about the systems which are time delay and dominating time constant. In this study, the system is perturbed with pseudo random binary sequence (PRBS) as an input and temperature as an output. PRBS was chosen as the input signal, since a correctly designed sequence is persistently exciting for any system. The input signal PRBS was generated for different possible proportional band and perform in short durations rather than one long session. All data were collected without sample in the steam tray and done by open loop behavior. The MATLAB was used to collect the input-output data.

4.3.1 Unit Conversion

Figure 4.7 [1] illustrates the Simulink model that was used in the unit conversion. The initial output signal measured in voltage form was converted to temperature in degree Celsius, (°C). This unit conversion is the second-order polynomial

Fig. 4.6 PLCF method

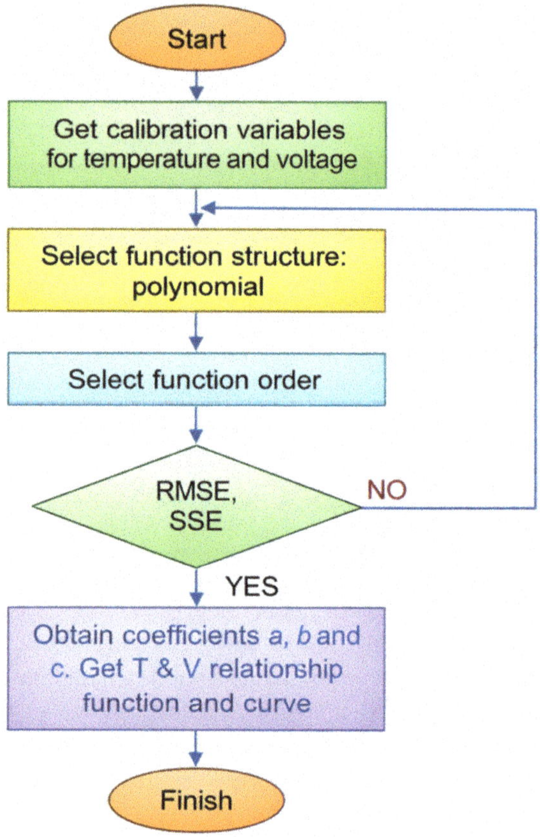

that was derived during the calibration procedure. The °C was set as a default temperature unit.

4.3.2 *The Step Test*

The step test of a system is the output of the system when a unit step function is used as the input. Typically, when a new system is designed, the step test of the system is the first characteristic of the system to be analyzed. This test was done by drawing a tangent line at the steepest point of the step response curve. From this analysis, it provides two information about the dynamic process of the system: (1) time constant, T_c and (2) time delay, T_d. The time constant is the time required for the output response to rise or fall exponentially through approximately 63 per cent of its amplitude as expressed in Eq. 4.4 [1]:

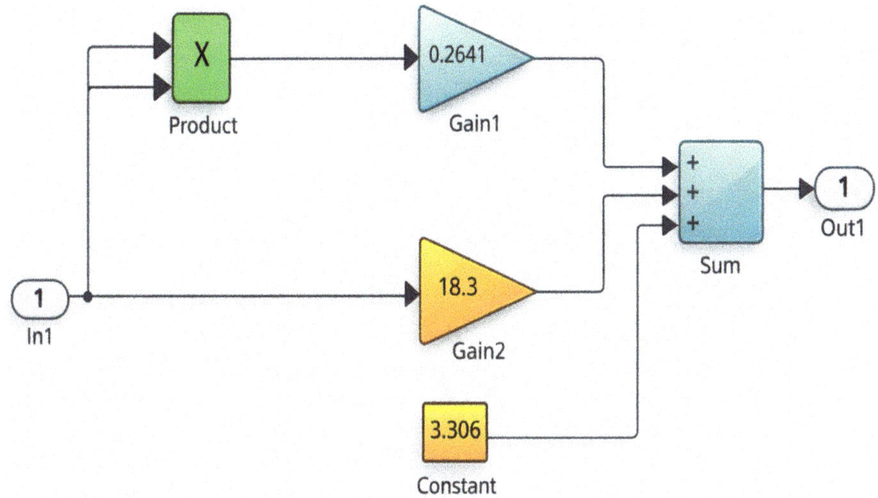

$$y = 0.2641x^2 + 18.3x + 3.306$$

Fig. 4.7 Simulink model for unit conversion

$$T_c \text{ temp} = \left[\frac{63.2}{100}(T_{\max} - T_{\min})\right] + T_{\min} \tag{4.3}$$

$$T_c = T_c \text{ time} - T_d \tag{4.4}$$

where T_d is the time taken by output starts to respond corresponding to the change in the input. T_d is estimated at the intersection between the baseline of the old steady-state level and the tangent line. The test was done with the 1 s sampling interval and at room temperature. The input signal is defined as an Eq. (4.5) [1]:

$$u(t) = \begin{cases} 1 & 1 \le t \le 5500 \\ 0 & \text{otherwise} \end{cases} \tag{4.5}$$

4.3.3 The PRBS Perturbation

The PRBS perturbation is a signal that shifts between two levels, which are switched to high or low values. In this work, the Matlab was used to generate the PRBS input by using an *id* input function in command line as below:

$$\begin{aligned} u &= idinput(N, \text{type, band, levels}) \\ u &= idinput\left(N, 'prbs', [0\ B], [0\ 1]\right) \end{aligned} \tag{4.6}$$

Table 4.1 Summary of PRBS properties

Number of PRBS sequences	N_A
PRBS clock period	$1/B$
N_{PRBS}	$N_A \times N_B$
PRBS * clock period (Δt)	$(1/B) \times N_B$

where u is the PRBS input sequence, N is the number of the generated input data, probability band, B where B will produce a signal that is constant over intervals of minimum length *1*/*B*, also known as the clock period. For instance, let say $B =$ 0.2, then the PRBS clock period 1/0.2 = 5 sequences. Whereas the *level's* function defines the input level such that the signal, u will always between the minimum value (0) and maximum value (1). However, the challenging part in designing the PRBS perturbation is the appropriate value of B that is most suitable for hydro-diffusion steam distillation system by affecting process dynamics under test. Different PRBS signals will excite different system dynamics. Table 4.1 tabulated the PRBS properties [1].

The N_B is called clock period factor, and it will determine the sampling intervals each PRBS sequence is sent to the system under test. So, the PRBS signals to use this N_B will be constant over intervals of $(1/B) \times N_B$. The Δt will influence the variation of process output. Therefore, in selecting the suitable value of Δt, the step test should be performed prior the PRBS tested [1].

The heuristic method was applied by conducting several experiments with a different B range from low perturbation, 0.02 to high perturbation, 1. At last, the PRBS signals with $B = 0.4$ were chosen as this signal able to affect the process dynamics.

The PRBS signals from PC will be sent to the electric heater through the power controller. The PRBS signal sent to the process is on/off signal, which drives the power controller to control the 1500 W electric heater. The sampling period was one data per second. The PRBS amplitude is selected to adhere at maximum of 100% of the full scale (5 V) when PRBS signals is 1 and 50% of full scale (2. 5 V) when the PRBS signals at 0. Figure 4.8 shows the flowchart to generate the PRBS data collection.

Flowchart to generate PRBS data collection [1]

See Fig. 4.8.

4.4 Linear ARX Model

In this section, system identification was used to build linear ARX model to describe the underlying mechanism of the observed data. The ARX modeling was done by using the Matlab command line. The organizational structure for this sub-chapter

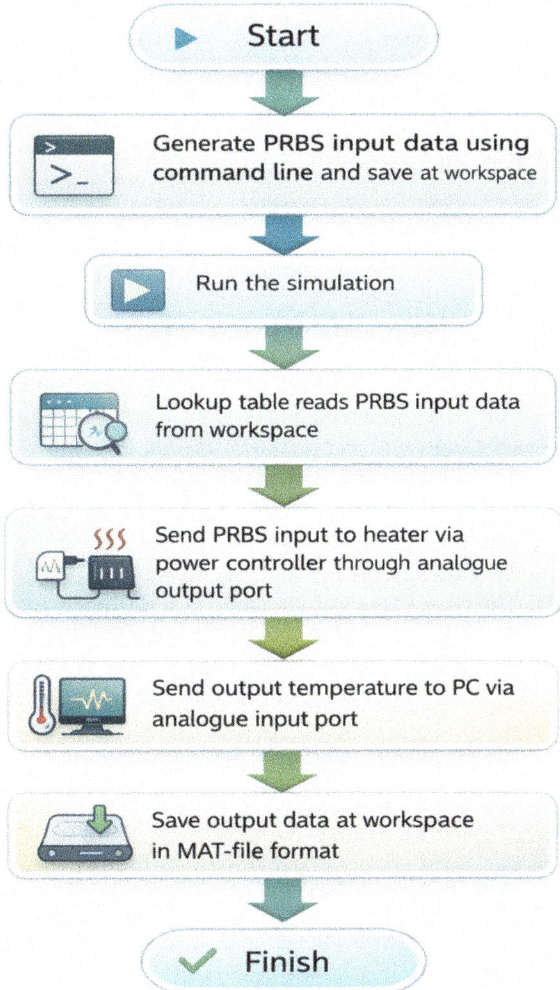

Fig. 4.8 Flowchart to generate PRBS data collection

is as follows: Sect. 4.4.1 is data organization; Sect. 4.4.2 is model order selection; Sect. 4.4.3 is model estimation and finally, Sect. 4.4.4 model validation.

4.4.1 Data Organization

There are 3 sets of data that will consider performing the ARX modeling namely as a data set Z_1, Z_2, and Z_3. Each data will represent different dynamics and different behavior. The steps taken in data organization as follows:

Table 4.2 Data organization

Data	Temperature	Model's name	Estimation data	Validation data
*z*1 (*u*, i)	From 70.8 to 94.1 °C (range 23.3 °C)	arx1	*z*1*ue* and *z*1*ye*	*z*1*uv* and *z*1*yv*
*z*2 (*u*, *y*)	From 76 to 92 °C (range 16 °C)	arx2	*z*2*ue* and *z*2*ye*	*z*2*uv* and *z*2*yv*
z3 (*u*, *y*)	From 79 to 85 °C (range 6 °C)	arx3	*z*3*ue* and *z*3*ye*	*z*3*uv* and *z*3*yv*

I. The modeling approach start with the collected PRBS input and steam temperature output that were saved in excel file were loaded into the Matlab workspace by using "import data" icon. The output steam temperature for data set Z_1, Z_2, and Z_3 are denoted as *z*1*y*, *z*2*y*, and *z*3*y*, respectively. Meanwhile for the input data represent as *z*1*u*, z2*u*, and *z*3*u*.
II. The data samples were preprocessing by divided equally into 2 sets: odd number sample's data (1, 3, 5 ...) for *z*1 (*u*, *y*), *z*2 (*u*, *y*) and *z*3 (*u*, *y*) are used for estimation, the even number samples (2, 4, 6 …) are used for validation purposed by using interlacing technique.
III. The *z*1 (*u*, *y*) *e*, *z*2 (*u*, *y*) *e*, and *z*3 (*u*, *y*) e are input-output estimation data and *z*1 (*u*, *y*)*v*, *z*2 (*u*, *y*)*v*, and *z*3 (*u*, *y*)*v* are input-output validation data. The arrangement of estimation data and validation data for the ARX model is tabulated as in Table 4.2 [1].

4.4.2 Model Order Selection

The model order is chosen as low as possible to generate the ARX model. In this study, ARX 111 was chosen and means that the model order for [*na, nb, nk*] equal to [1, 1, 1].

4.4.3 Model Estimation

The unknown parameters of the ARX model are estimated based on the linear regression method. This method is based on minimizing the criterion of loss function. The function *arx* in System Identification Toolbox was used to estimate the ARX model. The fitted ARX model is named as arx1, arx2, and arx3 for data *z*1, *z*2, and *z*3, accordingly.

4.4.4 Model Validation

The purpose of model validation is to validate the estimated ARX model for each data set using its validation data. The model validation is done based on visual inspection and statistical evaluation by using the percentage of fit, R^2, 1-SAP with residual, correlation test and lastly residual histogram. The *compare* function will compute the output in percentage of fit by comparing validation data with the estimated model.

The 1-SAP will compare the predicted data and validation data. The discrepancy between these two data is called residual or error. The residual will be measured using RMSE, and the good model will produce low RMSE value. At first, input-output validation data must be converted to "iddata" using *iddata* function.

The function *resid* is applied in MATLAB programming to perform the correlation analyses which are combined the auto-correlation test (ACF) and cross-correlation test (CCF) for each data set. The correlation test is to evaluate the whiteness of residual by using 95% confidence limit.

The residual histogram was used to check the residual distribution by visual inspection. The residual distribution will be checked using mean and variance. A good model should have the residual that is normally distributed. The command line used for residual histogram is using *hist* function.

4.5 Controller Design

In this study, three types of controllers will be concentrated, which are the conventional PID, HFPID and STFPID controllers. The purpose in designing a controller is regulating the steam temperature to the operating point to achieve some objectives such as robust, transient performances and optimizes the economic function. Simulations and real time implementation will be carried out. The effectiveness of the controller will be measured by using the performance indices of settling time, rise time; percentage overshoots (%OS), and RMSE.

4.5.1 Simulation Controller Design

The aim of the simulation study is to determine the preliminary results of the controller settings. The conventional PID, hybrid fuzzy plus PID, and self-tuning fuzzy PID controller were developed to perform the closed-loop system simulation based on the approximated ARX model. Then, the performances for all controllers will be compared.

Table 4.3 The PID parameters using Ziegler-Nichols method

Parameter	Value
K_p	0.013
K_i	0.0000271
K_d	31.2

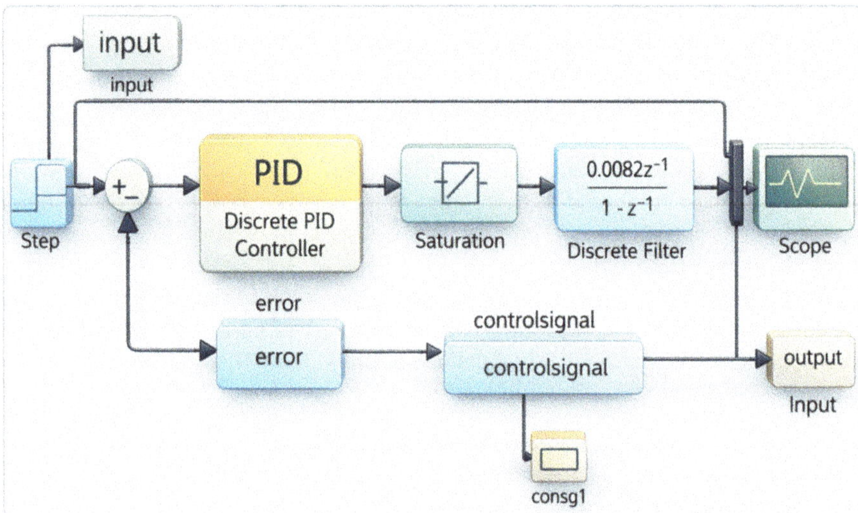

Fig. 4.9 Simulink block diagram for simulation of PID

4.5.1.1 PID Controller

The PID controller is a simple three-term controller where the parameters of proportional gain (K_p), integral gain (K_i) and derivative gain (K_d) were tuned using Ziegler-Nichols method. The PID parameters are tabulated in Table 4.3 [1]. However, fine tuning was applied to ensure an optimum result. The sampling period, T_s is 1 second.

Figure 4.9 presents the Simulink block diagram for PID controller [1]. The plant model arx1 was attached to the PID controller to control the steam temperature. The step input was set at the 90 °C as a reference input. The saturation block will limit the control signal to the plant and was set between 0 and 5 V. Meanwhile, the error is the difference between reference input and output.

4.5.1.2 Hybrid Fuzzy Plus PID Controller

A hybrid fuzzy plus PID (HFPID) controller is proposed that is integrated to hydro-diffusion steam distillation essential oil extraction process. The HFPID is designed with the objectives that there should be no overshoot in the output response, and shorter rise time to make the system faster. This is because the process is categorized as slow process and longtime delay. In this study, we integrate conventional PID and fuzzy logic controller to exploit the beneficial sides of both controllers. Figure 4.10 illustrates the Simulink of HFPID structure for controlling the steam temperature in hydro-diffusion steam distillation [1].

The HFPID controller is the combination of a fuzzy controller with PID controller where the structures are connected in cascade. This controller uses two inputs, which are error, (e) and derivative of error, $de(t)/dt$ and one output, $u(t)$. Equations (4.7) and (4.8) stated the algorithm for HFPID [1]:

$$n = F\{kp_1 e, Kd_1 \mathrm{d}e(t)/\mathrm{d}t\} \tag{4.7}$$

$$u = \Big((Gf_1 \cdot n)xGe + \big((Gf_2 \cdot n)xG_e \cdot K_{i_2} \textstyle\int_0^\tau e(\sigma)\mathrm{d}(\sigma)\big) + \big((Gf_3 \cdot n)x(Ge \cdot Kd_2 \cdot e(\tau))\big)\Big). \tag{4.8}$$

The control signal, u (0–5 V) to drive the heater for the hydro-diffusion system is calculated based on the summation of all products from fuzzy components and PID components. Whereas n signal is derived from fuzzy operation, $F\{\}$ that comprises of two inputs; error and derivative of error as stated in Eq. (4.7). When there are differences between the current temperature and target temperature, the HFPID will force the response to minimize the error signal by increase or decrease the control signal. Otherwise, if the error shows the zero, the controller setting will be maintained at present condition. The connection diagram of a fuzzy system is shown in Fig. 4.11 [1].

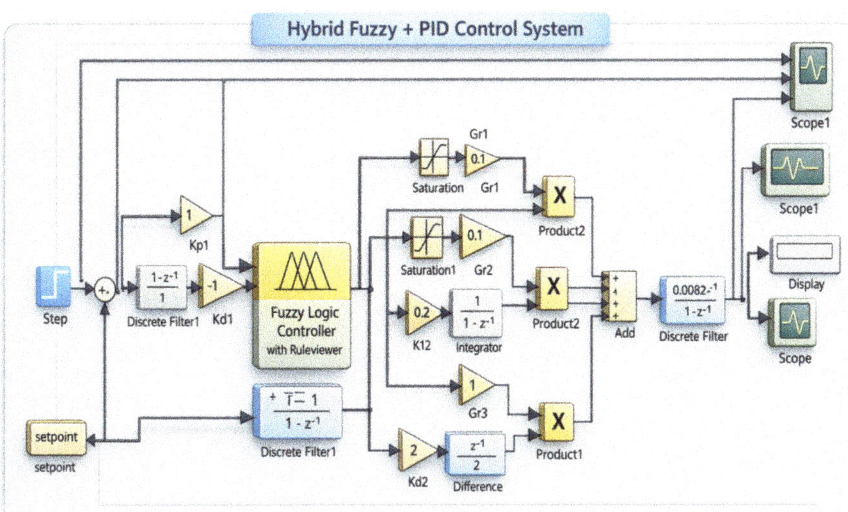

Fig. 4.10 Simulink block diagram for simulation of hybrid fuzzy plus PID

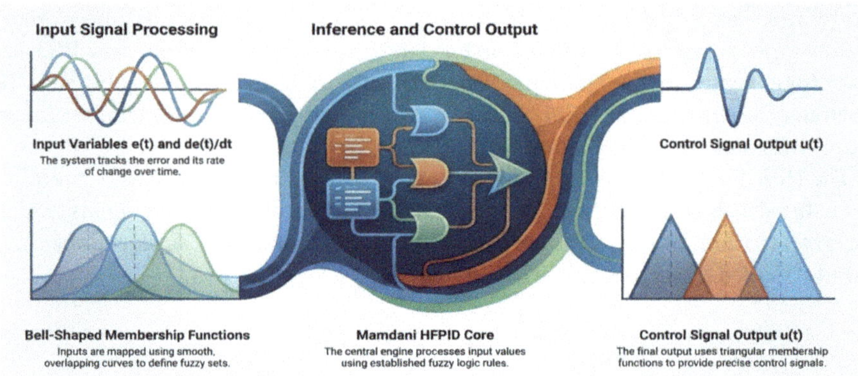

Fig. 4.11 The connection diagram of a fuzzy system in HFPID controller

Triangular and trapezoidal shape will be used in these works because of their simplicity and efficiency of implementation in the steam distillation process. The graphical representation using 7 membership functions for input and output variable are depicted in Figs. 4.12, 4.13, and 4.14 below [1].

The universe of discourse for LN was set at [− 100, − 100, − 75, − 50], MN was set at [− 75, − 50, − 25], SN was set at [− 50, − 25, 0], ZE was set at [− 25, 0, 25], SP was set at [0, 25, 50], MP was set at [25, 50, 75], and LP was set at [50, 75, 100, 100].

The universe of discourse for LN was set at [− 50, − 50, − 37.5, 25]; MN was set at [− 37.5, − 25, − 12.5], SN was set at [− 25, − 12.5, 0]; ZE was set at [− 12.5, 0, 12.5]; SP was set at [0, 12.5, 25]; MP was set at [12.5, 25, 37.5], and LP was set at [25, 37.5, 50, 50].

The universe of discourse for LN was set at [− 1, − 1, − 0.75, − 0.5], MN was set at [− 0.75, − 0.5, − 0.25], SN was set at [− 0.5, − 0.25, 0]; ZE was set at [−

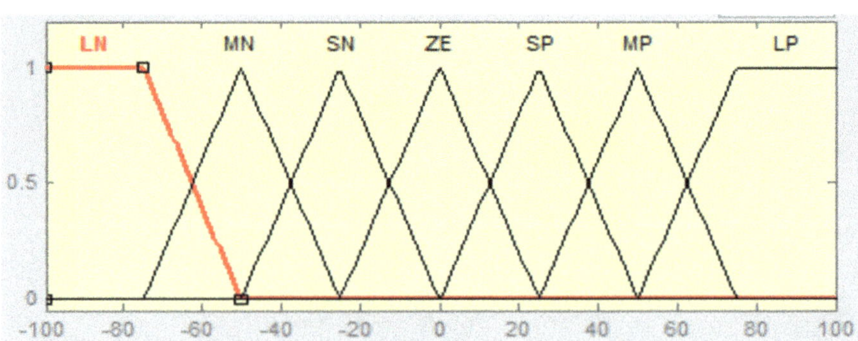

Fig. 4.12 Input error membership function

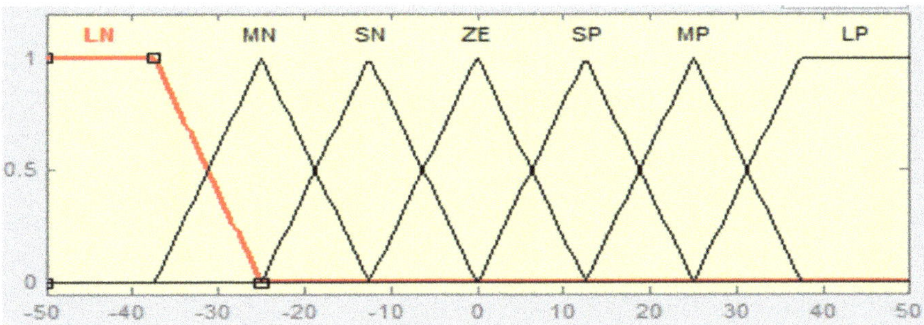

Fig. 4.13 Input derivative of error membership function

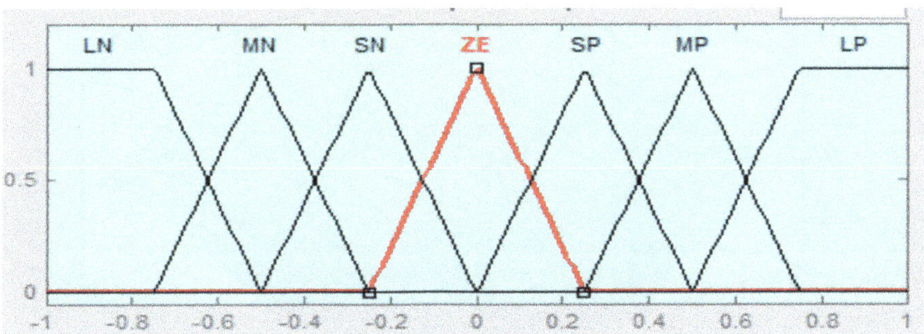

Fig. 4.14 Output membership function

0.25, 0, 0.25]; SP was set at [0, 0.25, 0.5]; MP was set at [0.25, 0.5, 0.75], and LP was set at [0.5, 0.75, 1, 1].

Table 4.4 represents linguistic rules for 7 membership functions, which contribute to 49 rules where LN is large negative, MN is medium negative; SN is small negative; ZE is zero; SP is small positive; MP is medium positive, and LP is large positive [1]. The linear control rules for HFPID controller are generated by:

- Rule: IF e_N is P_i and $\mathrm{d}e(t)/\mathrm{d}t_N$ is Q_i THEN u_{fz} is R_i

Figure 4.15 depicts the control surface for HFPID using 7 membership functions (HFPID-7) [1].

Simulation work already done to tune eight gain parameters in the HFPID as shown in Fig. 4.10. There are Kp_1, Kd_1, Gf_1, Gf_2 is fuzzy gains and Ge, Ki2; Kd2 are PID gains. All the parameters are tuned manually based on the knowledge of the process. Then the tuned structure of HFPID by simulation is tested on the real system. Table 4.5 tabulated the tuned parameter gains for HFPID [1].

Table 4.4 Linguistic rules using 7 membership functions for HFPID controller

$e(t)$ $de(t)/dt$	LN	MN	SN	ZE	SP	MP	LP
LN	SN	MN	LN	ZE	SN	MN	LN
MN	SN	MN	LN	SN	SN	MN	LN
SN	SN	MN	LN	SN	SN	MN	LN
ZE	LP	MP	MP	LP	MP	LP	LP
SP	MP	LP	MP	LP	LP	LP	LP
MP	MP	LP	MP	LP	LP	LP	LP
LP	MP	LP	MP	LP	LP	LP	LP

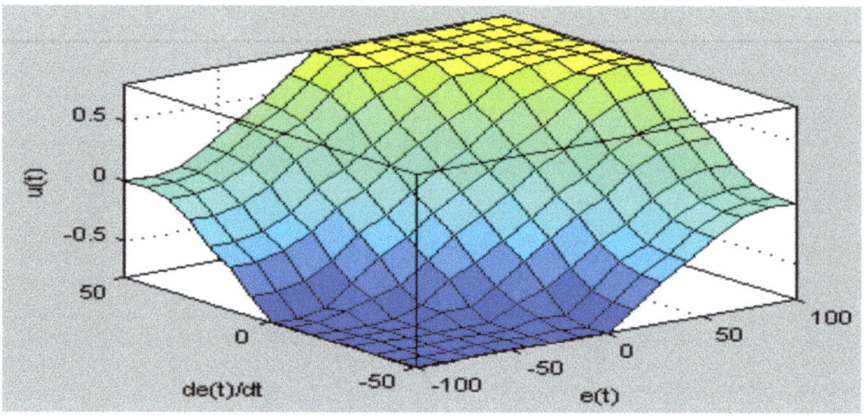

Fig. 4.15 The control surface for HFPID-7

Table 4.5 The Parameter's value of HFPID controller

Parameter	Gain
Kp_1	1
Kd_1	− 1
Gf_1	0.1
Gf_2	0.1
Ge	35
Ki_2	0.2
Kd_2	2

4.5.1.3 Self-tuning Fuzzy PID Controller

The self-tuning fuzzy PID (STFPID) controller is designed for the purpose to evaluate the behavior of this controller and to evaluate the degree of improvement given by STFPID for temperature control compared with conventional PID and HFPID controller. Figure 4.16 shows the Simulink representation of steam temperature control of hydro-diffusion steam distillation system using STFPID controller [1].

The reference setpoint was set at 90 °C. Firstly, two input variables must be defined in terms of linguistics form that is an error, $e(t)$ and derivative of error, d$e(t)$/$d(t)$. According to the current value of $e(t)$ and d$e(t)/d(t)$, fuzzy will automatically tune the three outputs of the PID parameters' controller which are K_p', K_i', and K_d'. Figure 4.17 depicts the connection diagram of the fuzzy system by implementing Mamdani-based fuzzy inference system [1].

In this study, the range of each PID parameter was determined based on the optimum fine tune in simulation of PID controller using ZN. So, the range of *Kp, Ki*, and *Kd* are $K_p \in [1, 10]$, $K_i \in [0.000001, 0.00001]$ and $K_d \in [10, 100]$, respectively. Therefore, they can be calibrated over the range [0, 1] as Eq. (4.9):

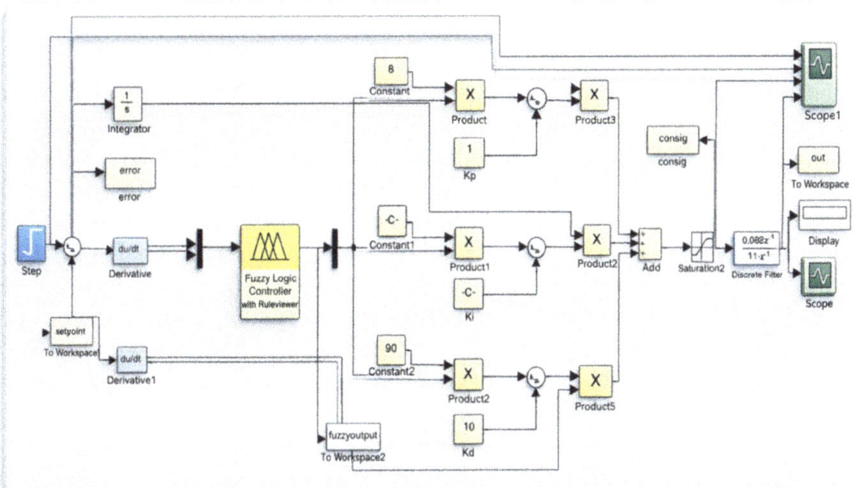

Fig. 4.16 Simulink block diagram for simulation of STFPID

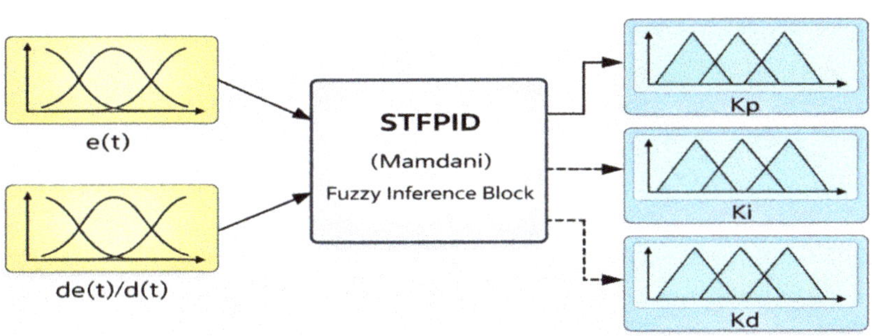

Fig. 4.17 The Mamdani fuzzy inference block

$$
\begin{aligned}
K_p' &= \frac{K_p - K_{p\min}}{K_{p\max} - K_{p\min}} \\
K_i' &= \frac{K_i - K_{i\min}}{K_{i\max} - K_{i\min}} \\
K_d' &= \frac{K_d - K_{d\min}}{K_{d\max} - K_{d\min}}
\end{aligned} \tag{4.9}
$$

Hence, the equation for all PID parameters as follows:

$$K_p = 9K'_p + 1 \tag{4.10}$$

$$K_i = 0.000009K_i' + 0.000001 \tag{4.11}$$

$$K_d = 90K_d' + 10. \tag{4.12}$$

The membership functions of the $e(t)$, $de(t)/d(t)$ and output K_p', K_i', and K_d' using 5 membership functions as shown in Figs. 4.18, 4.19 and 4.20 consecutively [1]. Referring to Fig. 4.18, there are five linguistic variables of error: negative big (NB), negative small (NS), zero (ZE), positive small (PS), and positive big (PB). The universe of discourse for error membership function was set between − 100 and 100 since the process was considered at operating point of 25–100 °C. Similarly, the fuzzy set of the derivative error and output signal is presented as [NB, NS, ZE, PS, PB] over the interval of − 10 to 10 and 0 to 1, respectively. All the linguistic variables were present by the combination of trapezoidal and triangular shape membership functions. The linguistic variables were set based on the characteristics and specification of hydro-diffusion steam distillation system.

Referring to Fig. 4.18, the universe of discourse for NB was set at [− 100, − 100, − 30, − 20], NS was set at [− 20, − 10, 0], ZE was set at [− 5, 0, 5], PS was set at [0, 10, 20] and PB was set at [20, 30, 100, 100].

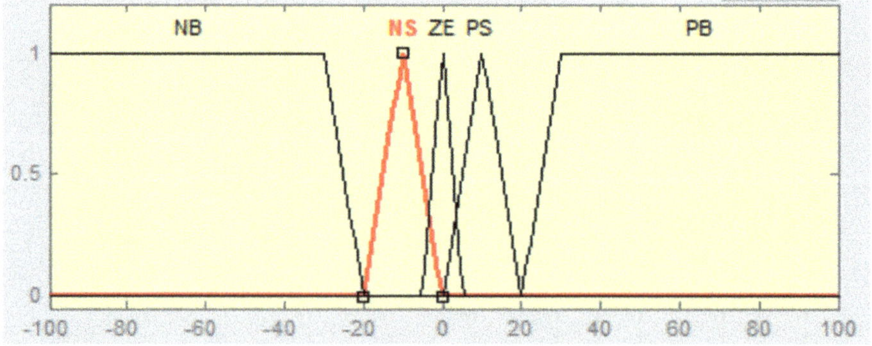

Fig. 4.18 Error membership function

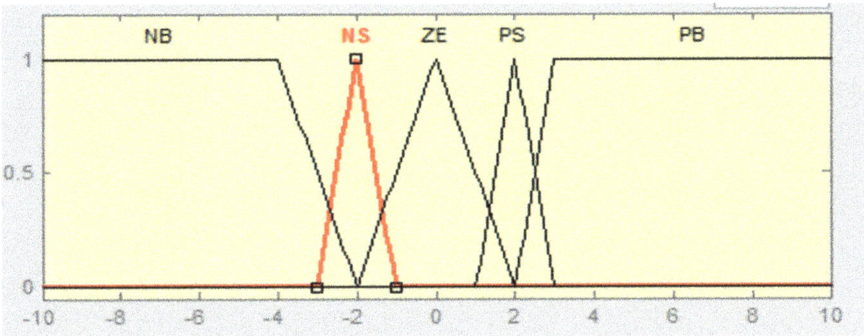

Fig. 4.19 Derivative of error membership function

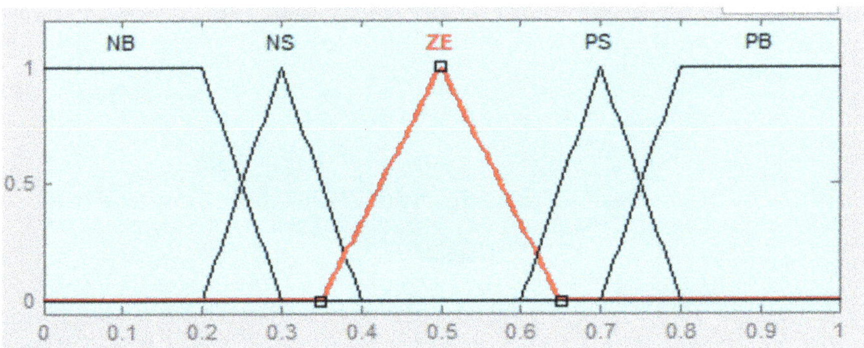

Fig. 4.20 KK'_p, K'_i, and K'_d membership function

As shown in Fig. 4.19, the universe of discourse for NB was set at [− 10, − 10, − 4, − 2], NS was set at [− 3, − 2, − 1], ZE was set at [− 2, 0, 2], PS was set at [1, 2, 3], and PB was set at [2, 3, 10, 10].

As can been seen from the Fig. 4.20, the universe of discourse for NB was set at [0, 0, 0.2, 0.3], NS was set at [0.2, 0.3, 0.4], ZE was set at [0.35, 0.5, 0.65], PS was set at [0.6, 0.7, 0.8], and PB was set at [0.7, 0.8, 1, 1].

The IF-THEN rules containing two antecedences, and one consequence were applied during computation algorithms. In this work, the rules' bases were built based on practical experience and knowledge about the plant to be controlled. The IF-THEN rules using 7 membership functions are summarized in Table 4.6 [1].

Hence, the fuzzy rule can be read using IF-THEN rule as below:

$$\text{If } e(t) \text{ is PB and } \frac{\mathrm{d}e(t)}{\mathrm{d}(t)} \text{ is PB THEN output is PB}$$

Table 4.6 The IF-THEN rules using 5 membership functions for STFPID controller

$e(t)$ $de(t)/dt$	NB	NS	ZE	PS	PB
NB	NB	NB	PS	PB	PB
NS	NB	NB	ZE	PB	PB
ZE	NB	NS	ZE	PS	PB
PS	NB	NB	ZE	PB	PB
PB	NB	NB	NS	PB	PB

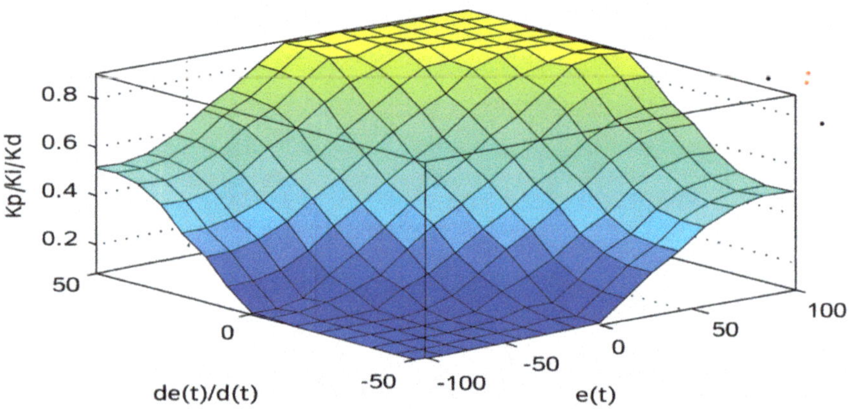

Fig. 4.21 The control surface for STFPID-5

The total numbers of rules are 25 rules since the input variables are 5 and 5 output variables. Figure 4.21 illustrated the control surface of self-tuning fuzzy PID controller using 7 membership functions (STFPID-5) [1].

4.5.2 Real-Time Implementation

The designed controllers were implementing to perform online control in hydro-diffusion steam distillation system after getting some inputs from simulation studies. Real-time implementation was done by using data-acquisition board (DAQ) PC LabCard Advantech PCI-1711 and by employing real-time window target (RTW) toolbox available in Matlab. This PCI-1711 card will be interfaced between hardware and control unit through port A1. The computer will be used as a control unit. The temperature control was achieved by controlling the voltage fed to the heater. The controller will continuously compare the desired set point with the actual output of steam temperature. Consequently, generate the adequate control signal. The controller output will be sent to the analog output and the analog input will generate the process output in voltage. The volt-temp block will be converting the process output in the voltage to the temperature in °C. The sampling time is 1 s per sample. Figures 4.22, 4.23, and 4.24 depicted the Simulink block diagram for real time PID, HFPID, and STFPID controller, consecutively [1]. The desired steam temperature was set at 90 °C. Again, the performances for all the controllers will be analyzed and compared.

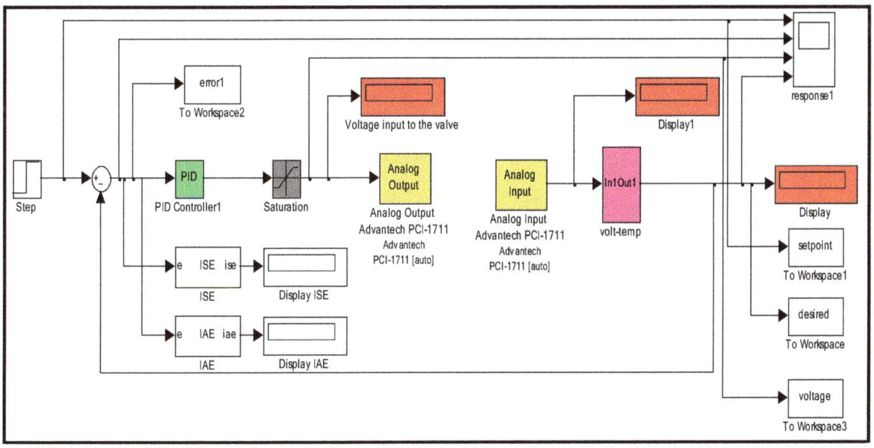

Fig. 4.22 Simulink block diagram for real time implementation of PID

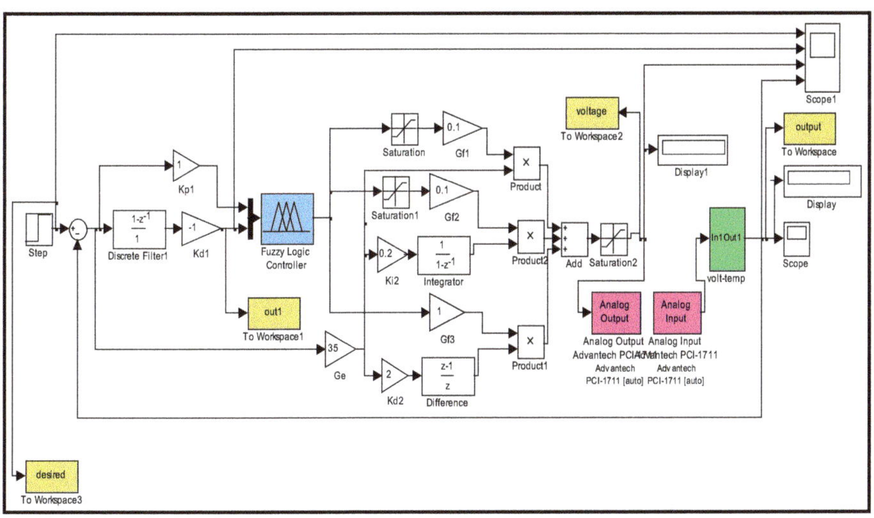

Fig. 4.23 Simulink block diagram for real time implementation of HFPID

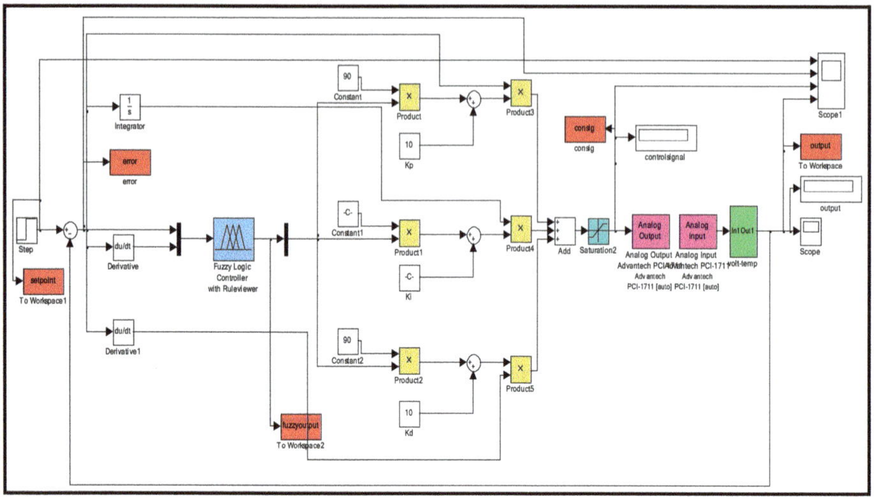

Fig. 4.24 Simulink block diagram for real-time implementation of STFPID

4.5.3 Robustness Test

To evaluate the robustness of the PID, HFPID, and STFPID controller in controlling steam temperature for extraction process, two robustness tests have been carried out via simulation and real time implementation. The tests are: (i) set point tracking and (ii) introduction of load disturbance.

(i) Set point tracking setting for simulation and real time implementation—The controller performance will be assessed based on the set point tracking where the output is expected to follow the reference signal, $r(k)$ with the objective to minimize the error between the output and desired response. The Simulink block diagram for set point tracking same as the step response, except the step input block change with the repeating sequence block. The performance analysis is based on rise time, settling time, %OS and RMSE.

$$r(k) = \begin{cases} 0^{\circ}\text{C} & \text{for } 0 < k < 2400 \\ 60^{\circ}\text{C} & \text{for } 2401 < k < 4400 \\ 80^{\circ}\text{C} & \text{for } 4401 < k < 6400 \\ 90^{\circ}\text{C} & \text{for } 6401 < k < 8400 \end{cases} \tag{4.13}$$

(ii) Load disturbance- For simulation purposed, the disturbance test was done by injecting 10 °C down from set point starting at 5000th samples. Robustness of the controller design will be based on how fast the output returns to the desired set point. For the real-time implementation, each controller will be introduced to the disturbance by cut off the power supply to the heater from 1.5 kW to 0 W for 30 s. It starts at sample 1000th until 1030th (during steady-state condition).

The aim of this test is to analysis how fast each controller can recover load disturbance during upset conditions. The performance comparison is based on the time required to adapt with load disturbance.

References

1. Yusoff, Z.M.: Self-tuning Fuzzy PID Control of Hydro-diffusion Essential Oil Extraction System. Universiti Teknologi MARA, Malaysia (2014)
2. Ozel, M.Z., Kaymaz, H.: Superheated water extraction, steam distillation and Soxhlet extraction of essential oils of *Origanum onites*. Anal. Bioanal. Chem. **379**(7–8), 1127–1133 (2004). https://doi.org/10.1007/s00216-004-2671-5
3. Peng, F., Sheng, L., Liu, B., Tong, H., Liu, S.: Comparison of different extraction methods: steam distillation, simultaneous distillation and extraction and headspace co-distillation, used for the analysis of the volatile components in aged flue-cured tobacco leaves. J. Chromatogr. A. **1040**(1) (2004). https://doi.org/10.1016/j.chroma.2004.03.057
4. Bousbia, N., Abert Vian, M., Ferhat, M.A., Petitcolas, E., Meklati, B.Y., Chemat, F.: Comparison of two isolation methods for essential oil from rosemary leaves: hydrodistillation and microwave hydrodiffusion and gravity. Food Chem. **114**(1), 355–362 (2009). https://doi.org/10.1016/j.foodchem.2008.09.106
5. Masango, P.: Cleaner production of essential oils by steam distillation. J. Clean. Prod. **13**(8), 833–839 (2005). https://doi.org/10.1016/j.jclepro.2004.02.039
6. Mahabob, N.Z., Yusoff, Z.M., Ismail, N., Taib, M.N.: Improving the quality of agarwood oil by using the intelligent of multilayer perceptron (MLP) network training algorithms. In: AIP Conference Proceedings (2023). https://doi.org/10.1063/5.0115568
7. Vian, M.A., Fernandez, X., Visinoni, F., Chemat, F.: Microwave hydrodiffusion and gravity, a new technique for extraction of essential oils. J. Chromatogr. A. **1190**(1–2), 14–17 (2008). https://doi.org/10.1016/j.chroma.2008.02.08

Chapter 5
Decoding the Steam Distillation Process: Prototype, Calibration, and Modeling

Abstract This chapter presents the implementation and analysis of a hydro-diffusion steam distillation system for essential oil extraction, focusing on prototype development, calibration, data acquisition, and system modeling. A laboratory-scale experimental setup is developed, incorporating key components such as a distillation column, material tray, condenser, heating element, temperature sensors, and a data acquisition system for real-time monitoring and control. The system is designed to enable accurate measurement of steam temperature and facilitate controlled experimentation. Calibration of the temperature measurement system is performed to establish the relationship between voltage and temperature, ensuring accurate data acquisition. Experimental data are collected using both step input and pseudo-random binary sequence (PRBS) signals to capture the dynamic behavior of the system under different operating conditions. The collected data sets are then used to develop linear auto-regressive with exogenous input (ARX) models representing the relationship between input power and output temperature. Model validation is conducted using multiple performance indicators, including best fit, coefficient of determination (R^2), root mean square error (RMSE), mean square error (MSE), and residual analysis. Additional validation using autocorrelation (ACF) and cross-correlation (CCF) tests is performed to assess model adequacy. Among the developed models, the selected model demonstrates high prediction accuracy and satisfactory performance in both transient and steady-state conditions.

5.1 Introduction

This chapter will be divided into 4 main sub-chapters start with the result for prototype development includes plant verification, calibration result, and data collection. The result from the ARX modeling is presented and the validation results for ARX modeling are based on percentage of best fit, R^2, 1-SAP, RMSE, histogram, ACF, and CCF. It will be followed by sub-topic which is controller performance results that will evaluate and validated in terms of rise time; percentage overshoots, settling time, and RMSE.

Z. Mohd Yusoff et al., *Fuzzy Logic, Smart Control, and the Future of Essential Oils*, Studies in Fuzziness and Soft Computing 442,
https://doi.org/10.1007/978-981-92-2084-7_5

5.2 The Hydro-diffusion Steam Distillation Essential Oil Extraction System

The photo of hydro-diffusion steam distillation essential oil extraction system is depicted as in Fig. 5.1 and installed at distributed control system laboratory (DCS) at Faculty of Electrical Engineering, University Technology Mara Shah Alam, Malaysia [1].

Figure 5.2 shows the schematic diagram of a hydro-diffusion system apparatus [1]. The automated hydro-diffusion steam distillation system provides flexibility where a broader spectrum of materials can be extracted such as kaffir lime, serai wangi, pandan and ginger. It's consisted of two main parts, which is the distillation column itself and material tray. The experimental device composed of distillation column, electric heater type is immersion heating element, a pipe, material tray, condenser, and container vase. Two thermocouples, RTD PT-100 measure the steam temperature in the middle of the material tray and water temperature in the distillation column were used.

Fig. 5.1 The hydro-diffusion steam distillation essential oil extraction system

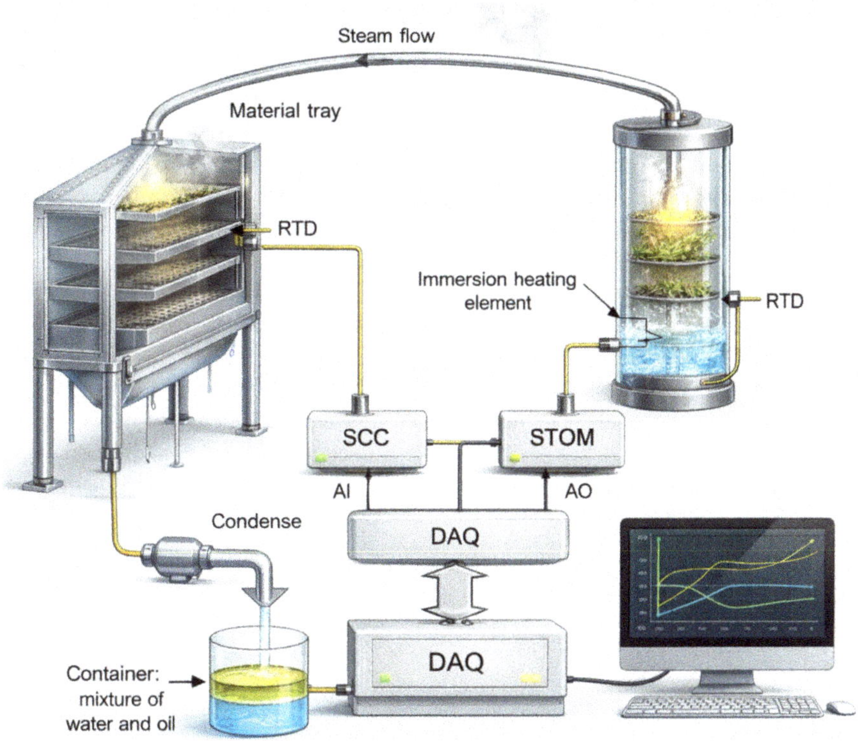

Fig. 5.2 The schematic diagram for complete system set-up

The distillation column was made using stainless steel because it offers many advantages such as high corrosion resistance allowing it to be used in rigorous environments, it is resistance to fire and heat allowing it to resist scaling and retain strength at high temperatures. An immersion heating element was located at the bottom of the distillation column responsible to heat the water. This type of electrical heater is better compared to gas and boiler because less expensive, quick to respond and the temperature can be controlled easily. Based on the steam rate, which is 1.5 l per hours, thus it was built to accommodate up to a maximum 10 l water and can support extraction process for almost 3 h. The calculation of steam rate was done by condensing the steam then measures the volume of the condensed water per hour. This step is important to avoid dry boiling. The special feature of this column is it was built by attaching to the self-refilling water line. The function of the self-refilling water line is to sustain and fixed the water level in the column as desired by compensating the evaporated steam in the distillation column. The warmer coolant water that initially used in condenser will feed-in from the bottom of the column without disturbing the water to steam phase at the top of the water surface. This is because the difference temperature of water flow in and inside the column will reduce and automatically decrease the column temperature disturbance. This special characteristic is important to produce accurate result, and for system identification purposed as compared with the conventional steam distillation system. The water flow-in is 2 l per hours and is determined based on the steam rate calculation. Figure 5.3 depicts self-refilling water line in hydro-diffusion steam distillation system [1].

A 50 cm length × 2 cm diameter pipe is used to connect the distillation column and material tray. The pipe act as a transportation medium to supply the steam produces in the distillation column to material tray. The material tray was built using stainless steel. The dimensions of the material tray are 12 inches × 12 inches × 40 inches height. This tray was equipped with 3 drawers, 1 inch's stainless-steel tube at the top and 6 mm stainless steel tube at the right and left side. The tube was designed in such a way to distribute consistently the steam for all space in the material tray. The opening door was design at the side to provide large opening for loading and unloading the raw materials. The raw material tray is then attached to the condenser and collecting container. A copper condenser is used to cool and condense the vapor of the volatile compound. The isometric view of the material tray is illustrated in Fig. 5.4 [1].

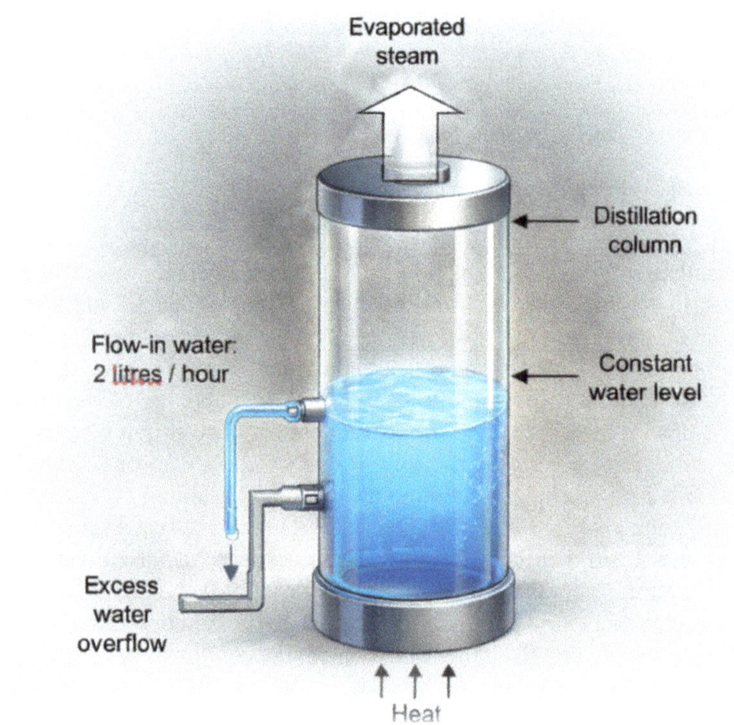

Fig. 5.3 The self-refilling water line

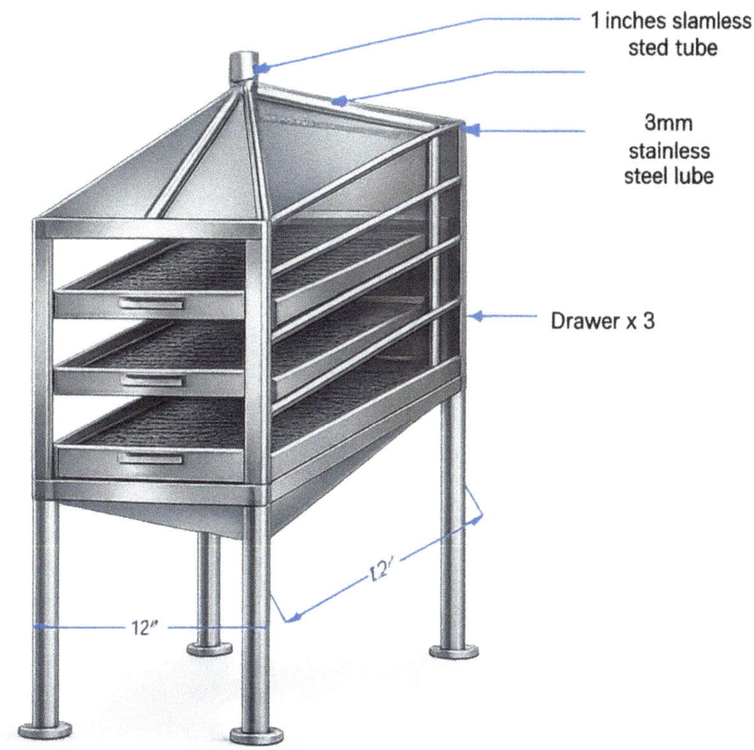

Fig. 5.4 The material tray

The extraction of essential oil process starts by placing the fresh or dry botanical materials in the material trays. The distillation column with 10 l water is heated by immersion heater to produce steam. Then, pressurized steam from distillation column will supply to the top of the material trays and passed through the botanical materials to vaporize the volatile oils in the botanical materials and release them. Material tray was attached to a condenser build from copper material and water oil separator. The steam and oil vapor mixture are then going down through the small tube to the condenser. Two layers form after the extraction process where oil on top of container and hydrosol at the bottom. Then, the volatile components were collected in a container by using a separator funnel.

System configuration consists of three main parts, which are system plant, data acquisition using PCI-1711 card, and lastly control unit. The block diagram is shown in Fig. 5.5 [1].

Computer was used as a control unit. DAQ PCI-1711 Advantech Lab Card was used to interface between hardware and control unit through port AI1. The system consists of three phases 240 Vac immersion type heater with a power rating of 1. 5 kW. The sensor is a platinum sensing element, RTD PT-100 3 wires type placed

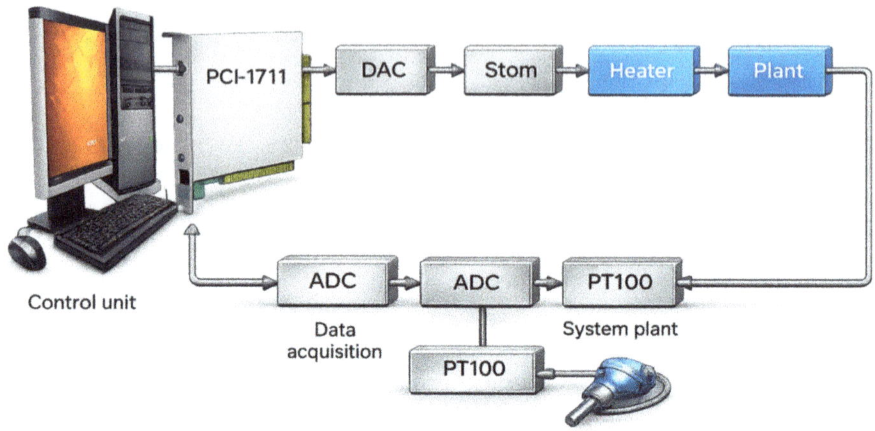

Fig. 5.5 Block diagram of system configuration

in the middle of the material tray to measure the steam temperature. These sensors are selected because they offer excellent accuracy, inexpensive, user-friendly, and wide range of applications. RTD PT-100 was connected to the signal conditioning circuit (SENECA K109PT) varies in terms of resistance value of 100 Ω at 0 °C. The power controller was regulating by steam to produce the voltage range from 0 to 5 V then send to the heater which acts as an actuator for the plant. Software for the system was developed using MATLAB version R2009a programming to monitor the signal response. The collected data at the workspace will be saving in. mat format. MATLAB Real-time Workshop (RTW) was employed for real-time implementation. The system specification was tabulated in Table 5.1 [1].

5.2.1 The Voltage-Temperature Calibration Curve

Table 5.2 tabulated the temperature and voltage data obtained from voltage-temperature calibration [1].

The fitted voltage-temperature calibration curve for first-order and second-order polynomial were depict in Figs. 5.6 and 5.7, respectively [1].

Then, the temperature relationship obtained as below, and the goodness of fit was tabulated in Table 5.3 [1].

For first-order polynomial:

$$T = 20.13v + 0.2548$$

For second-order polynomial:

$$T = 0.2641V^2 + 18.3V + 3.306$$

Table 5.1 System specification

Item	Specification
Power	1.5 kW, 240Vac, 50 Hz
Dimensions of	
1. Distillation column	Height = 65 cm, diameter = 22 cm, thickness = 1.4 mm
2. Condenser	Length = 50 cm, diameter_{in} = 1.5 cm, diameter_{out} = 2.8 cm
3. Material tray	Height = 14 inches, length = 12 inches, width = 12 inches
Materials	
1. Distillation column	Stainless steel
2. Condenser	Copper
3. Material tray	Stainless steel
4. Pipe	Heavy duty
Water volumes and rates	
1. Total volume of boiling water	10 l
2. Minimum water volume	3.5 l
3. Steam rate	1.5 l/h
Operation limit (time/volume)	3 h/6 l
Safety precautionary measures	Prevent dry boiling through self-refilling water line
Process monitoring	Water temperature in distillation column and steam temperature in the material tray

Table 5.2 Calibration result

Temperature (°C)	Voltage (V)
50	2.473
55	2.709
60	2.973
65	3.211
70	3.469
75	3.726
80	3.974
85	4.207
90	4.446

From the result, second-order polynomial was chosen as it produces a better result, more precise and more accurate compared with the first-order polynomial based on R^2, RMSE, and SSE.

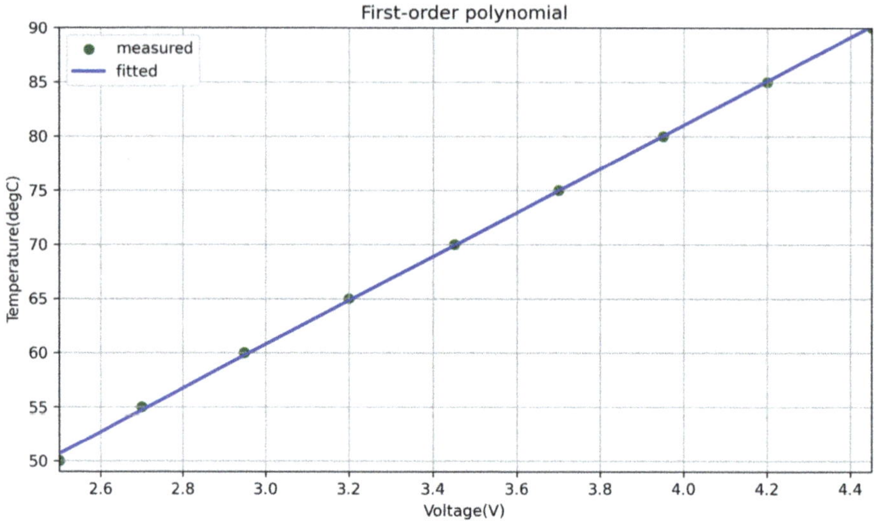

Fig. 5.6 The voltage-temperature calibration curve for first-order polynomial

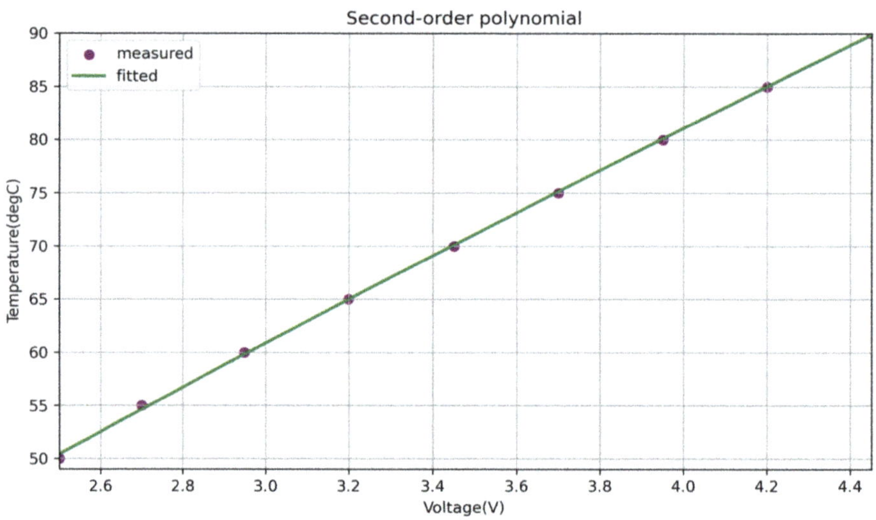

Fig. 5.7 The voltage-temperature calibration curve for second-order polynomial

Table 5.3 Goodness of fit

Order of polynomial	R^2	RMSE	SSE
1st	0.9998	0.1962	0.2695
2nd	0.9999	0.1781	0.1903

5.2.2 The Step Test

Figures 5.8 and 5.9 show the step input that was applied to the system and reaction curve method that applied on the output of the step test with 5500 samples of data collection, respectively [1].

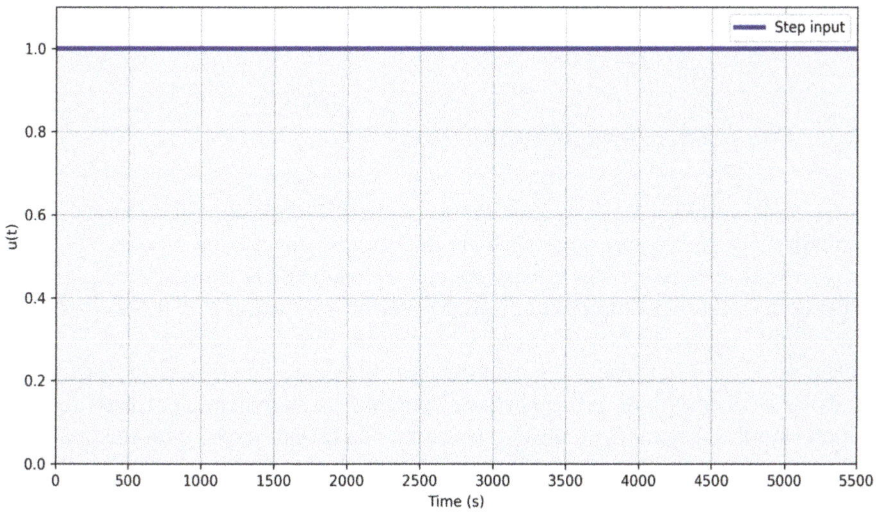

Fig. 5.8 The step input

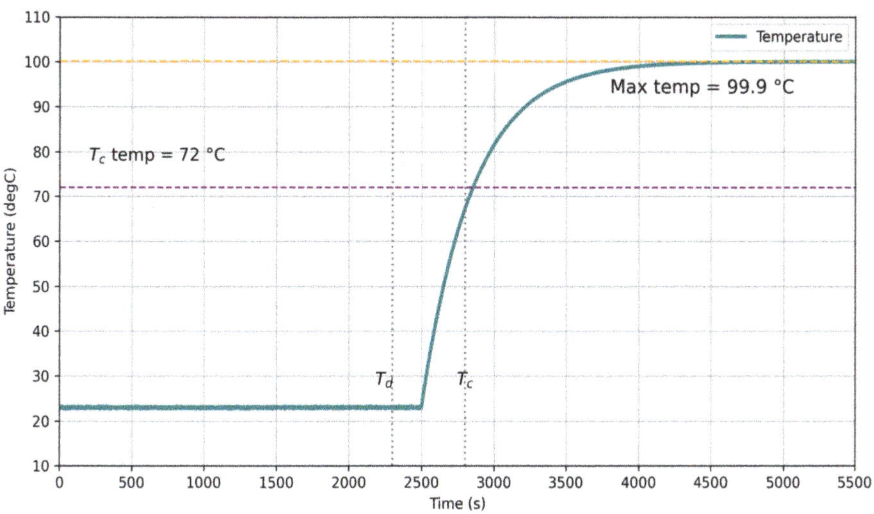

Fig. 5.9 Output of step test

The initial temperature at the steam tray is 23.31 °C and was gradually increased up to a boiling point of 99.9 °C. The time delay and dominating time constant are 2400 seconds and 400 seconds at T_c temperature 72 °C, respectively. The gain is 0.7679. Time delay for hydro-diffusion steam distillation system is longer than the conventional system because temperature starts increase when the steam start produces in the distillation column. When the temperature in the distillation column reaches 99.9 °C, then the temperature in the steam tray starts to increase and lastly achieves 99.9 °C.

5.2.3 The PRBS Input-Output Data

Figures 5.10 until Fig. 5.12 depicted the three set of PRBS input and output data of the steam temperature measured by RTD that will be considered in work [1]. Each data represents different system dynamics and were labeled as data set Z_1, Z_2, and Z_3. All the collected input-output data from heating process using $B = 0.4$ was denoted as: $u(t)$: input signal, PRBS (0, 1) and $y(t)$: output signal, steam temperature (°C).

Data set Z_2 in Fig. 5.11 is the data that was collected under feed-in disturbance based on PRBS perturbation [1]. The input-output data from heating process consists of 3000 samples. Figure 5.11 shows the output of the data set Z_2 with the operating range varies between 76 and 92 °C (range 16). It shows that the initial temperature

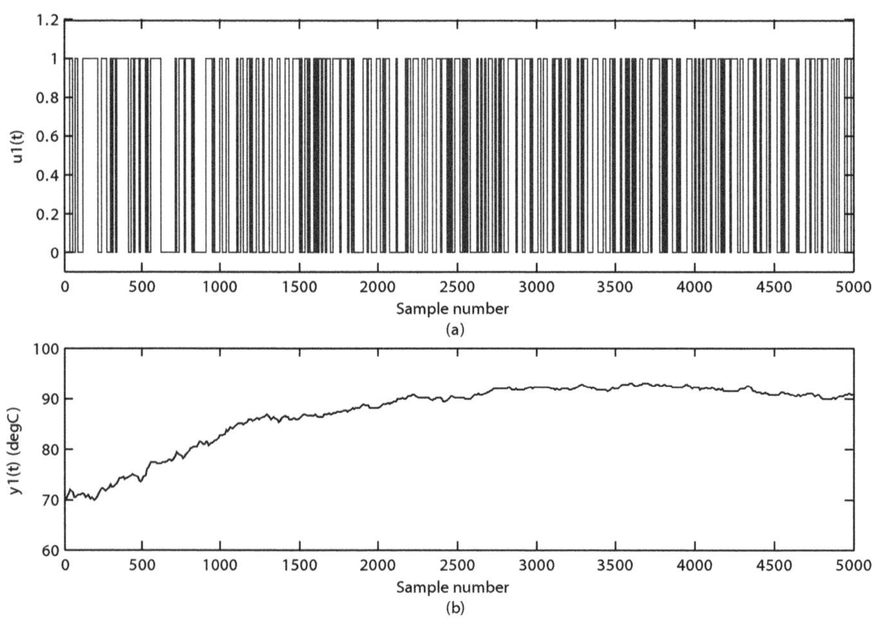

Fig. 5.10 PRBS data set Z_1. **a** Input, **b** output

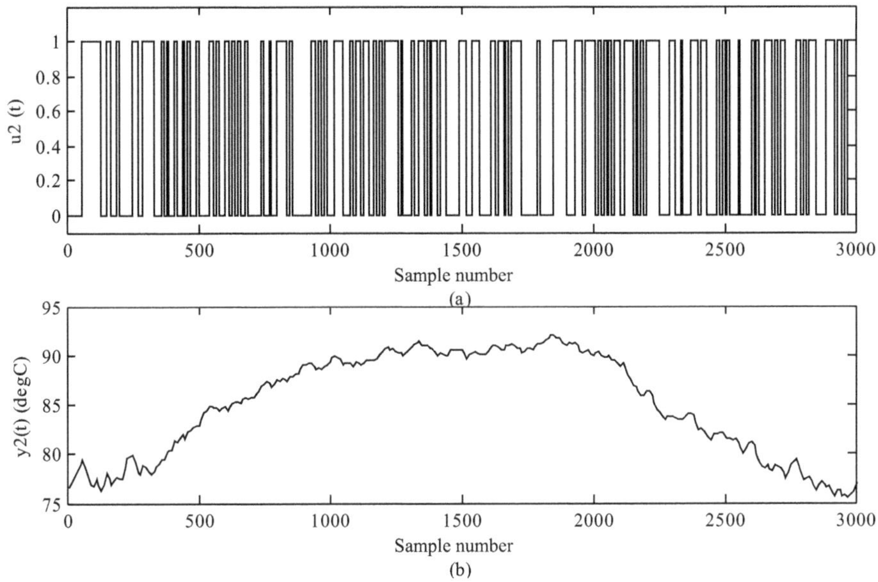

Fig. 5.11 PRBS data set Z_2. **a** Input, **b** output

is 76 °C and increase up to 92 °C but after 1934th sample's number, the steam temperatures start decrease. This situation because of feed-in disturbance applies on the distillation column. This technique applied to maintain the water level in the distillation column. However, the water feed-in which is 3 l per hour will disturb the water-to-steam phase where the temperature at the water surface will be significantly reduced, and vaporization process will be upset.

The input-output data set Z_3 from heating process consists of 3000 samples number. Figure 5.12 shows the output data with the narrowest operating range vary between 79 and 85 °C (range 6) [1]. From the figure, it clearly indicates that the measured output is totally depending on PRBS input switching. For example, the output variation between 500th and 1000th samples number shows PRBS input switching influence the measured output to vary.

5.3 ARX Model

This sub-section will discuss briefly on the result produce to obtain the three linear ARX models to present the hydro-diffusion steam distillation system. The estimated models that are based on the loss function criterion were assessing by performed 1-SAP and correlation's tests where the performance indices used to evaluate the result from 1-SAP such as R^2, RMSE, and residual histogram that comprise of mean and variance. The ACF and CCF tests were evaluated based on correlation function

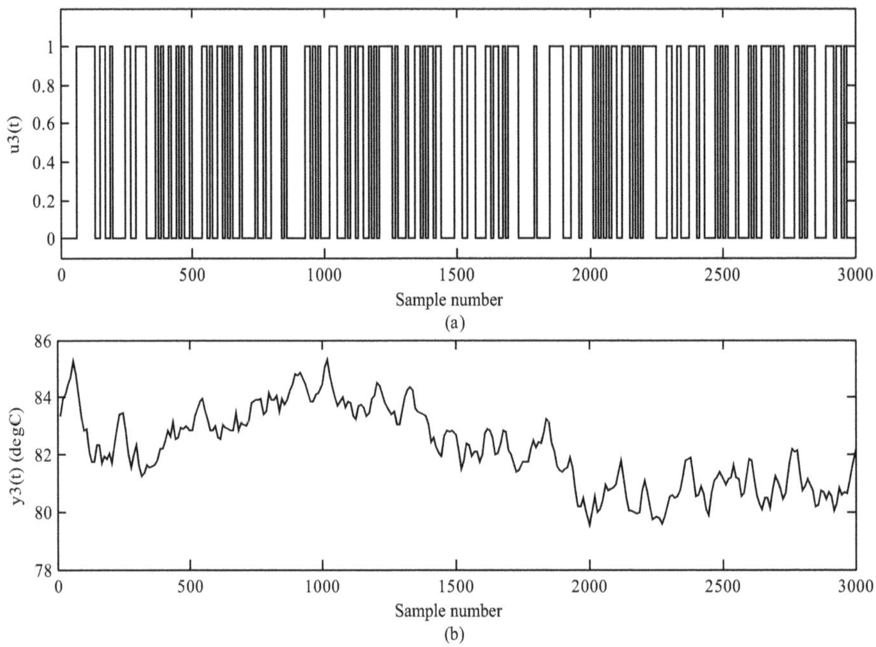

Fig. 5.12 PRBS data set Z_3. **a** Input, **b** output

between residuals and input signal. At the end of this sub-section, the well-estimated ARX model will be selected to perform further for controller design where a good model should be efficient in all conditions of the system.

5.3.1 Data Separation

Figures 5.13 and 5.14 show the estimation data and validation data extracted from data Z_1 [1]. Estimation data and validation data will have 2500 sample numbers. Since the data, Z_1 is sampled using 1 s sampling interval, so then the (u1e y1e) and (u1v y1v) will have 2 s sampling interval.

Figures 5.15 and 5.16 present the estimation data and validation data extracted from data Z_2 [1]. Estimation data and validation data will have 1500 sample numbers. Since the data, Z_2 is sampled using 1 second sampling interval, so then the (u2e y2e) and (u2v y2v) will have 2 seconds sampling interval.

Figures 5.17 and 5.18 illustrated the estimation data and validation data extracted from data Z_3 [1]. Estimation data and validation data will have 1500 sample numbers. Since the data, Z_3 is sampled using 1 second sampling interval, so then the (u3e y3e) and (u3v y3v) will have 2 seconds sampling interval, similar as data Z_1 and data Z_2.

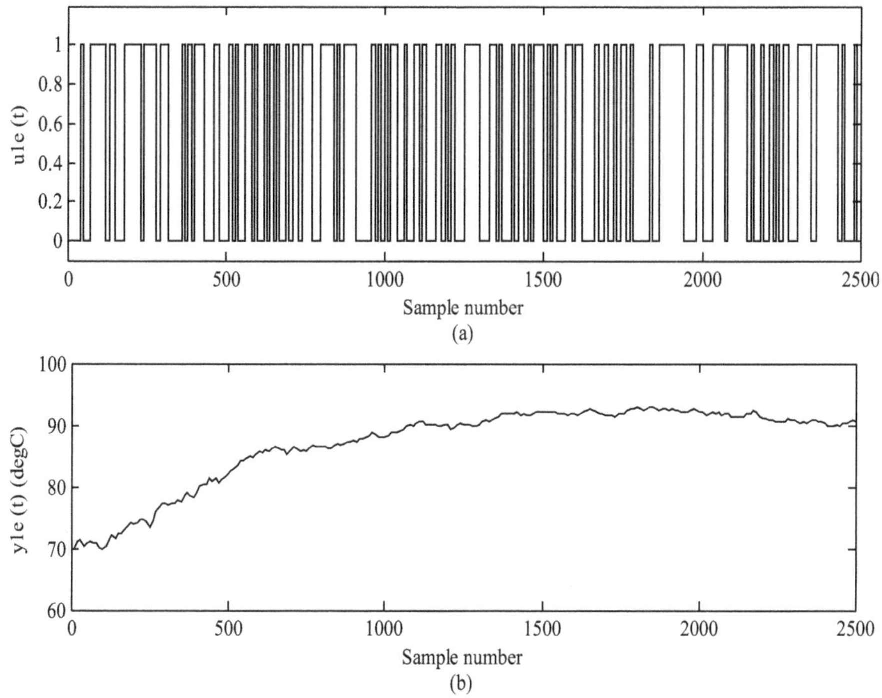

Fig. 5.13 Estimation data for Z_1. **a** Input, **b** output

5.3.2 ARX Model for Hydro-diffusion Steam Distillation System

The modeling results obtained from the data set Z_1, Z_2, and Z_3 are presented in Table 5.4 with the model's name as arx1, arx2, and arx3, respectively. Included in the Table 5.4 are transfer functions for each model [1]. The transfer functions are developed using the '*tf*' function provided by Matlab programming. The transfer function is discrete transfer functions that implement the z-transform transfer function, for design controller purposed. The transfer function is numerator, $B(q)$ over denominator, $A(q)$. One of this transfer function will be selected to b be used further in the controller design.

Then, the simulated ARX models in Table 5.4 were rearranged to represent in equation form for easier interpretation and observation as can be seen from Table 5.5 [1].

From the data in Table 5.5, it is apparent that the estimated ARX models from data set Z_1, Z_2, and Z_3 are first-order model, which are as expected. Its means that the order for polynomial $A(q)$ and $B(q)$ is equal to one. Further analysis showed that all the models namely as arx1, arx2, and arx3 relate the output and input by linear equation where the output, $y(t)$ is dependent on one previous output, one previous

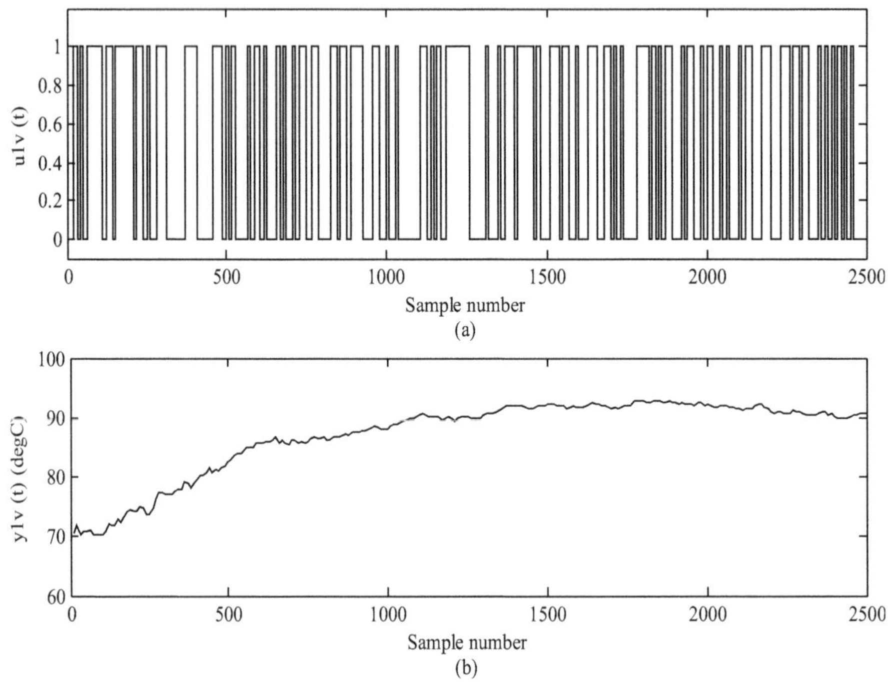

Fig. 5.14 Validation data for Z_1. **a** Input, **b** output

input with the one delay time response. There was a significant difference between all the models in the value of input numerator, $u(t)$ which is 0.0082, 0.01227, and 0.00478 for model arx1, arx2, and arx3, respectively. However, there was no difference in the value for output denominator, $y(t)$ where it's equal for all the models, which is one.

The estimated ARX models were compared in terms of loss function criterion as shown in Table 5.6 [1]. The most striking result to emerge from the data is that although all the models use the same model order which is one, but the loss function produce is different. This is because the sets of data are different in terms of interested system dynamic and their operating range. Model's arx1 produce smallest loss function, which is 0.01047, followed by model's arx2 with 0.01389, then lastly model's arx3 with the loss function 0.01534. A good model is the model that can provide minimum loss function.

5.3.3 1-SAP and Residual Histogram Model Validation

Figure 5.19 shows the measured output, 1-SAP output and residual plot for arx1 model validation [1]. The finding shows that the predicted output is in good agreement with the measured output. This is support by the residual plot where *residual* =

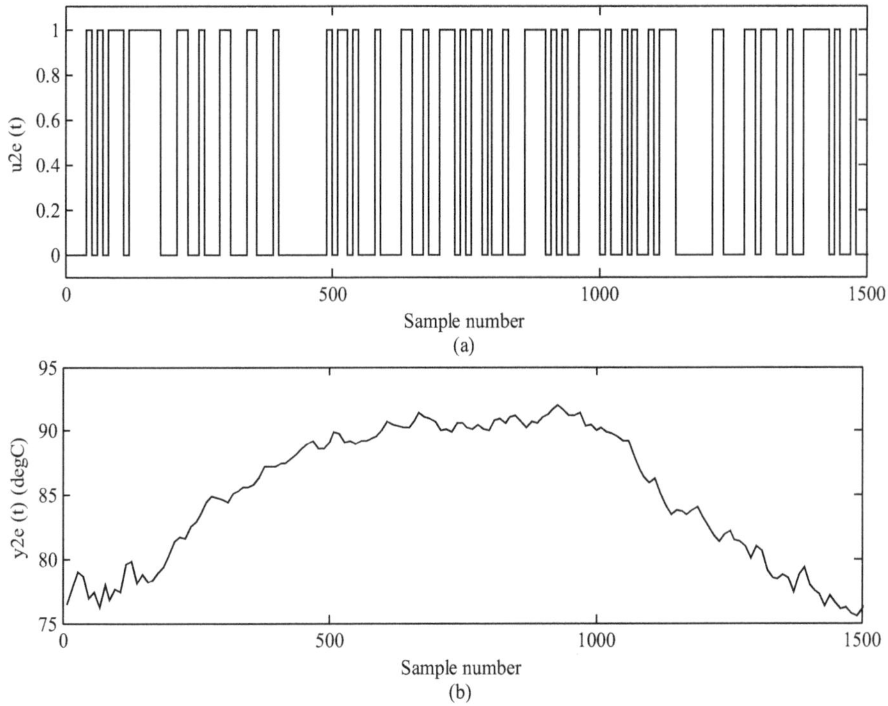

Fig. 5.15 Estimation data for Z_2. **a** Input, **b** output

predicted output—measured output. The best fit, R^2 and calculated RMSE, MSE for model arx1 are 98.42%, 0.1009, and 0.0102, respectively. Figure 5.20 depicted the residual histogram for model arx1 [1]. As can be seen from the figure, the residual is normally distributed with very small residual variance and residual mean. The residual mean is 0.0029 and residual variance is 0.0077.

Figure 5.21 shows the measured output, 1-SAP output and residual plot for arx2 model validation [1]. It is apparent in the figure that the predicted output is in good agreement with the measured output, which is confirmed by the residual plot. The best fit, R^2 and calculated RMSE, MSE for model arx2 are 97.68%, 0.1206, and 0.0145, accordingly. Figure 5.22 depicted the residual histogram for model arx2 [1]. The figure shows that the residual is normally distributed with very small residual variance and residual mean. The residual mean is 0.0023 and residual variance is 0.0086.

Figure 5.23 shows the measured output, 1-SAP output and residual plot for arx3 model validation [1]. It is apparent in the figure that the predicted output is in good agreement with the measured output, which is confirmed by the residual plot. The best fit, R^2 and calculated RMSE, MSE for model arx3 are 91.20%, 0.1214 and 0.0147, accordingly. Figure 5.24 illustrated the residual histogram for model arx3 [1]. From the figure, it is shown that the residual is normally distributed with very

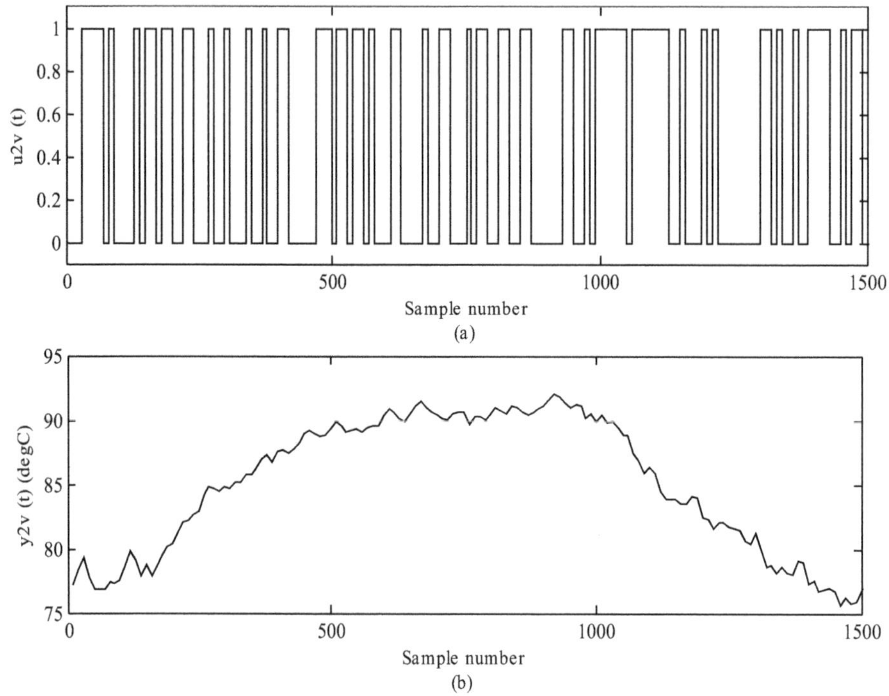

Fig. 5.16 Validation data for Z_2. **a** Input, **b** output

small residual variance and residual mean. The residual mean and residual variance are 0.0007934 and 0.0096, respectively.

Table 5.7 summarized performance indices for 1-SAP model validation which included best fit, R^2, RMSE, and MSE [1]. The residual distribution is analyzed by plotting the histogram then measured its mean and variance.

Figure 5.25 depicted the graphical comparisons between model arx1, arx2, and arx3 based on best fit, R^2 [1]. Referring to R^2 as shown in Fig. 5.25 and first column in Table 5.7, generally the results show that the best fit for all the models are very good, which are above 90% [1]. However, comparing of the result reveals that notable performance of arx1 by predicted the measured data with the highest best fit, 98.42%, then followed by arx2 (97.86%). Meanwhile, arx3 produce the lowest model fitted among three models, which is 91.2%. Greater significant differences were found between arx1 and arx3.

Figure 5.26 presents the graphical comparisons of RMSE and MSE between model arx1, arx2, and arx3, respectively [1]. From the figure and RMSE, MSE data that were tabulated in Table 5.7, it can be observed that arx1 consistently produced the best result as compared with arx2 and arx3 [1]. The RMSE and MSE results reveal that arx1 produce smallest values, followed by arx2 and arx3 consecutively. The good model should produce small RMSE and MSE.

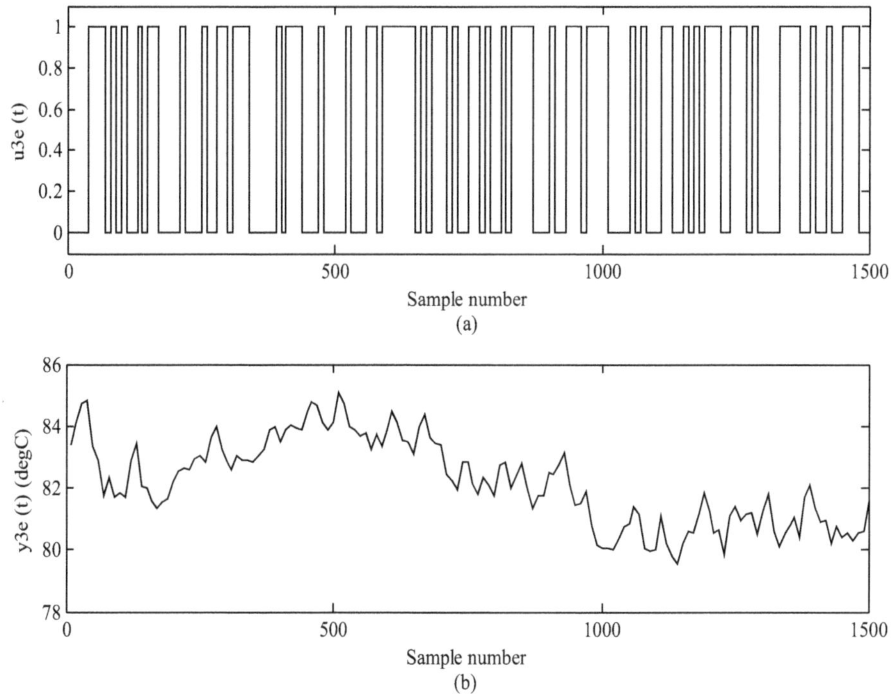

Fig. 5.17 Estimation data for Z_3. **a** Input, **b** output

Figure 5.27 depicted the graphical comparisons of mean and variance between model arx1, arx2, and arx3 [1]. The results clearly indicated that the residual histograms for all the models are normally distributed because mean and variance approximate to zero. From the Fig. 5.27 and fifth column in Table 5.7, it is obvious that arx3 produces the lowest significant residual mean, 0.0007934 as compared with arx1, 0.0029, and arx2, 0.0023. This situation is opposite to other performance indices. On the other hand, arx1 is slightly better and outperformed the arx2 and arx3 by producing smallest residual variance. Overall, based on ranking, arx1 is the most preferred model because produced smallest R^2, RMSE, MSE, and residual variance except residual mean.

5.3.4 ACF Test and CCF Test Model Validation

Figure 5.28 presents the correlation tests result for model arx1. The correlation test comprises of ACF test of the residual and CCF test between input residual for Figs. 5.28a, b, respectively [1]. The dashed line in both figures indicated the 95% confidence limit. The result as shown in Fig. 5.28a indicated that the residual has failed the whiteness test since some of the correlation coefficients are detected

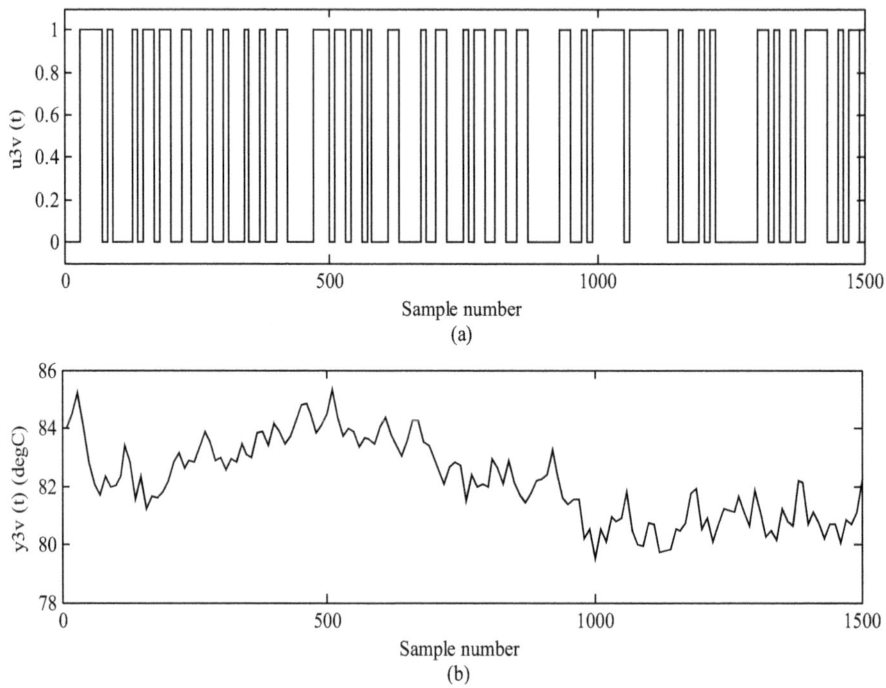

Fig. 5.18 Validation data For Z_3. **a** Input, **b** output

Table 5.4 ARX model for hydro diffusion steam distillation system

Model's name	ARX model	Transfer function
*arx*1	$A(q) = 1 - q^{-1}$ $B(q)0.0082q^{-1}$	$arx1 = \frac{0.0082z^{-1}}{1-z^{-1}}$
*arx*2	$A(q) = 1 - q^{-1}$ $B(q)0.01227q^{-1}$	$arx1 = \frac{0.01227z^{-1}}{1-z^{-1}}$
*arx*3	$A(q) = 1 - q^{-1}$ $B(q)0.00478q^{-1}B(q) = 0.00478q^{-1}$	$arx1 = \frac{0.00478z^{-1}}{1-z^{-1}}$

Table 5.5 Simulated general ARX model

Model's name	ARX model
arx1	$y(t) + y(t-1) = 0.0082u(t-1) + e(t)$
arx2	$y(t) + y(t-1) = 0.01227u(t-1) + e(t)$
arx3	$y(t) + y(t-1) = 0.00478u(t-1) + e(t)$

Table 5.6 Loss function for ARX model

Model's name	Loss function
arx1	0.01047
arx2	0.01389
arx3	0.01534

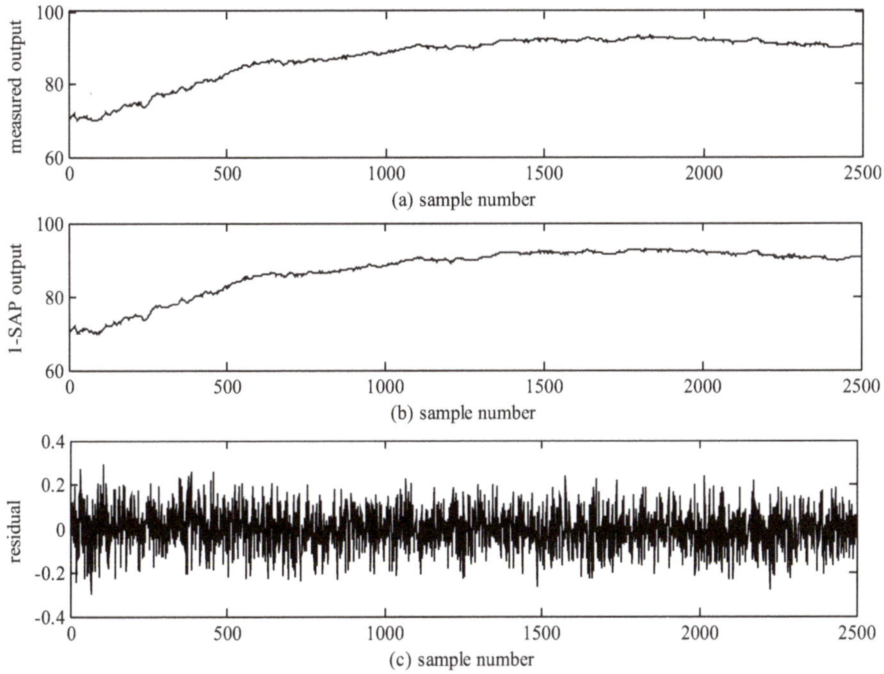

Fig. 5.19 Validation result for model arx1 using 1-SAP. **a** Measured output, **b** 1-SAP output, **c** residual plot

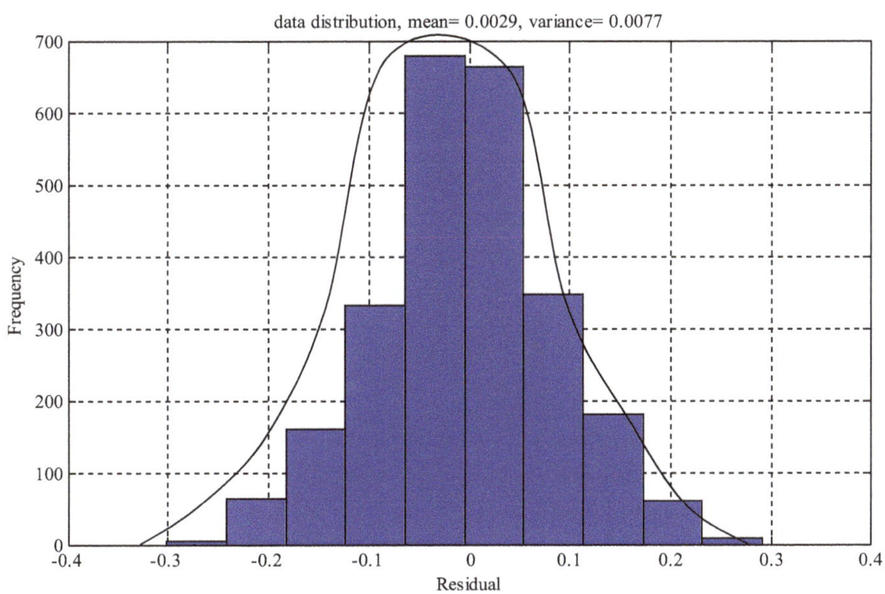

Fig. 5.20 Residual histogram for model arx1

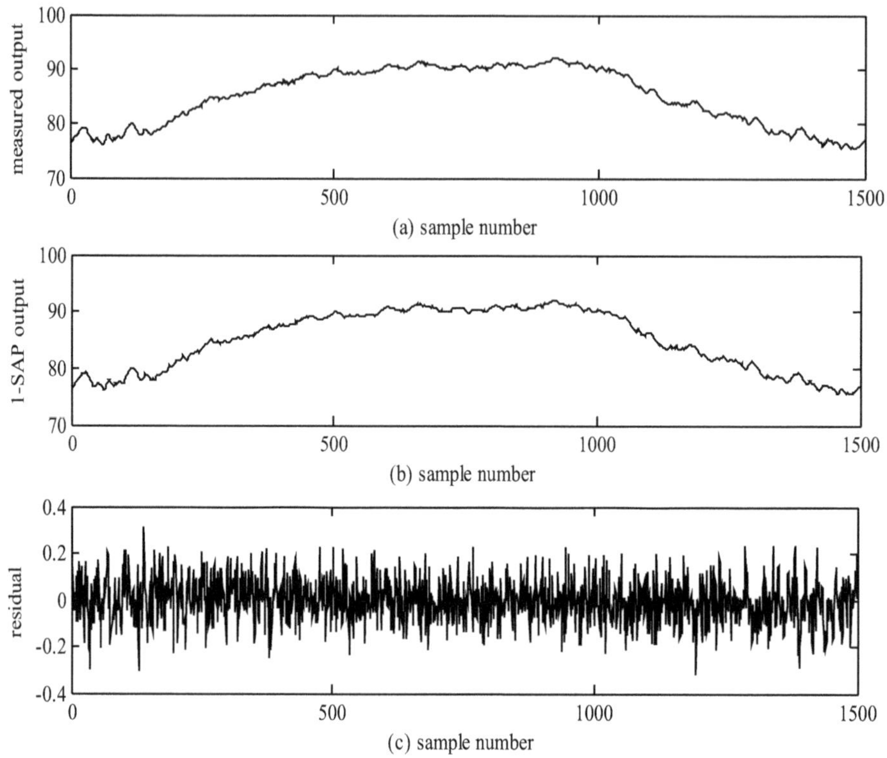

Fig. 5.21 Validation result for model arx2 using 1-SAP. **a** Measured output, **b** 1-SAP output, **c** residual

outside confidence limit. The correlation was detected at lags 1, 2, 3, 4, and 5. On the other hand, CCF test showed the five correlation coefficients were detected between input and residual, which are between lags 3 and 7.

Figure 5.29 depicted the ACF test and CCF test for model arx2. The ACF test of the residual and CCF test between inputs residual is shown in Fig. 5.29a and b, respectively [1]. The dashed line in both figures indicated that the 95% confidence limit. Referring to Fig. 5.29a indicated that the residual is correlated and has failed the whiteness test since some of the correlation coefficients are detected outside confidence limit. The correlation was detected at lags 1, 2, 3, 4, and 5. Meanwhile for CCF test, it's shown that there is a correlation between input and residual since some correlation coefficients lie outside confidence limit at lags 2, 3, 4, 5, 6, and 7.

Figure 5.30 shown the correlation tests result for model arx3. This test is comprised of ACF test of the residual and CCF test between input residual as present in Fig. 5.30a and b, respectively [1]. The dashed line in both figures indicated that the 95% confidence limit. The result as shown in Fig. 5.30a indicated that the residual is correlated since some of the correlation coefficients are detected outside confidence limit. The residual has failed the whiteness test. The correlation was detected at lags 1, 2, 3,

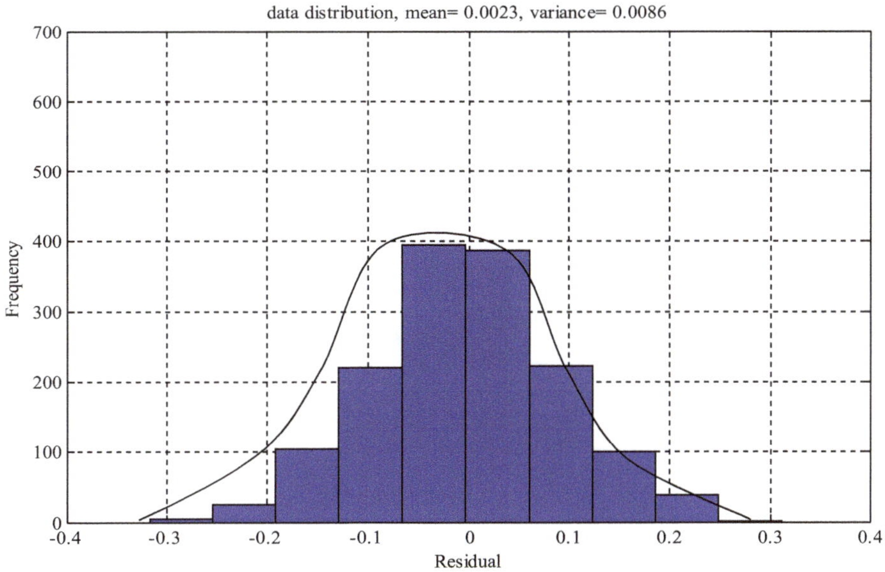

Fig. 5.22 Residual histogram for model arx2

and 4. On the other hand, CCF test showed the seven correlation coefficients were detected between input and residual, which are between lags 2 and 8.

Table 5.8 tabulated the summary of correlation tests for model arx1, arx2 and arx3 [1]. From the table, the ACF and CCF result for all models shows that the residual is correlated since some coefficients beyond the confidence limit. Its mean that the whiteness test for all models is lower since some information is not fully utilized by developed model structures. Referring to ACF test, there were no significant differences between model arx1 and model arx2 since both detect five correlation coefficients. Meanwhile, model arx3 shows lowest correlation detected, which are four. On the other hand, CCF test shows model arx1 detected five correlations between input PRBS and residual, followed by model arx2 with six correlation coefficients and lastly model arx3 with seven correlation coefficients. Although there are some correlations, coefficients were detected, but it can be accepted.

The present findings seem to be consistent with another study which found that ARX with first-order model is enough to present and describe the heating system of the essential oil extraction system. The model arx1 was chosen for further use in controller design since this model produces a better result by considering best fit, R^2, 1-SAP, RMSE, residual variance, residual mean, and correlation test. This model also covers transient state and steady-state condition, which are important for real system performances.

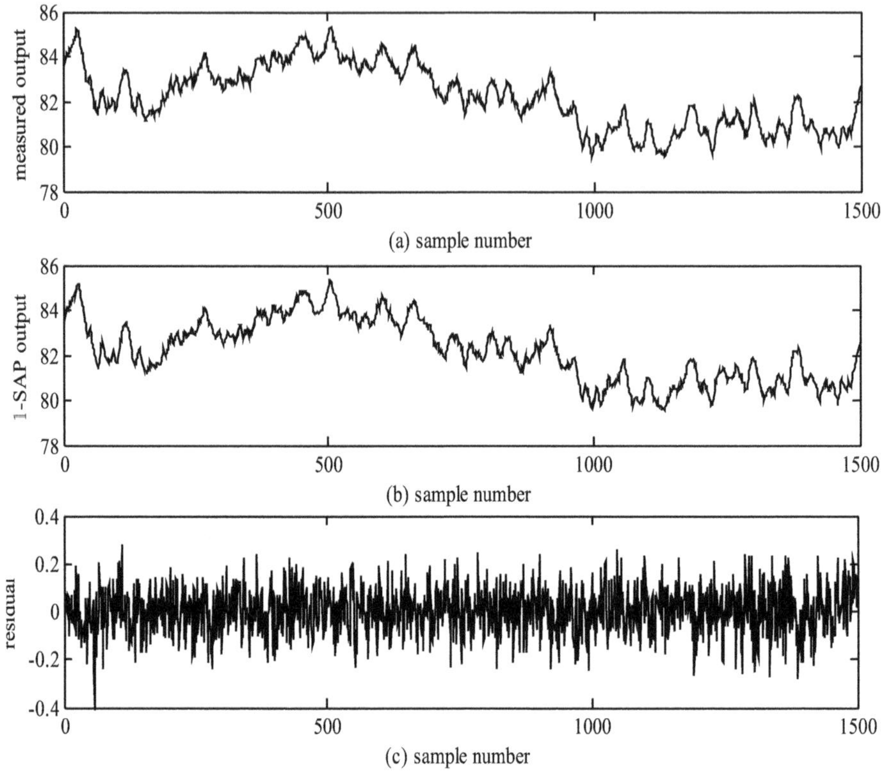

Fig. 5.23 Validation result for model arx3 using 1-SAP. **a** Measured output, **b** 1-SAP output, **c** residual plot

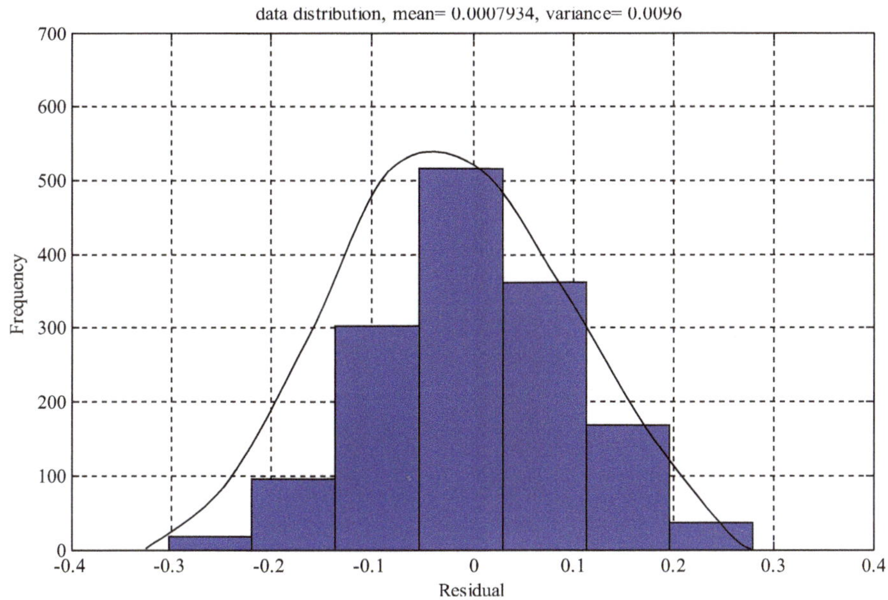

Fig. 5.24 Residual histogram for model arx3

Table 5.7 Performance indices for 1-SAP model validation

Model's name	Best fit, R^2	RMSE	MSE	Mean	Variance
arx1	98.42	0.1009	0.0102	0.0029	0.0077
arx2	97.68	0.1206	0.0145	0.0023	0.0086
arx3	91.2	0.1214	0.0147	0.0007934	0.0096

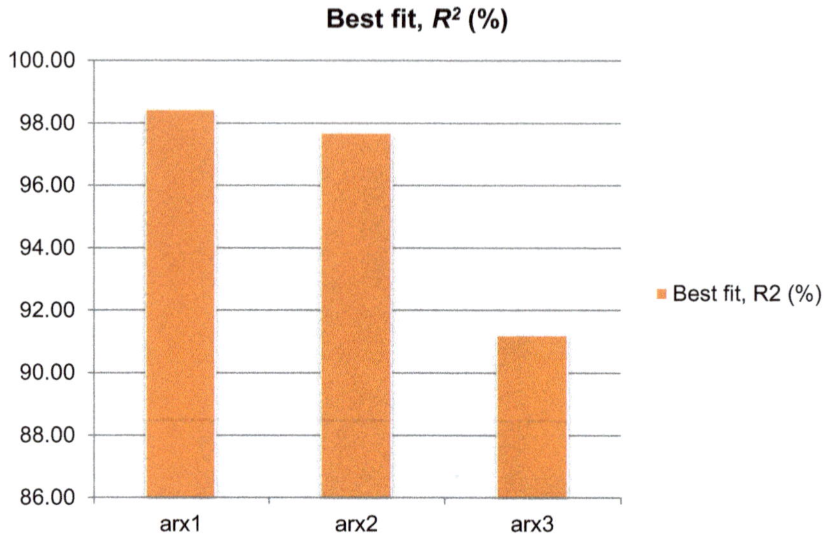

Fig. 5.25 Comparison of best fit, R^2 between model arx1, arx2, and arx3

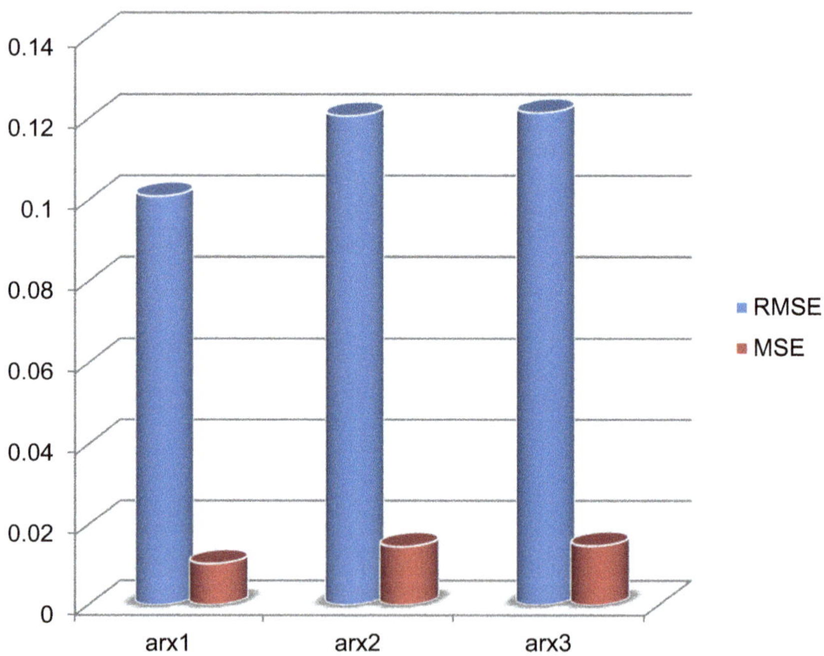

Fig. 5.26 Comparisons of RMSE and MSE between model arx1, arx2, and arx3

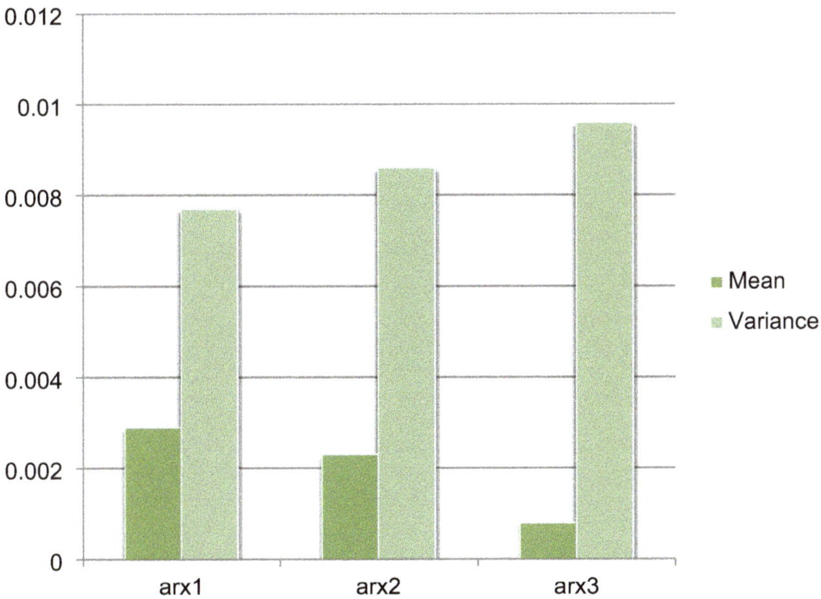

Fig. 5.27 Comparisons of mean and variance between model arx1, arx2, and arx3

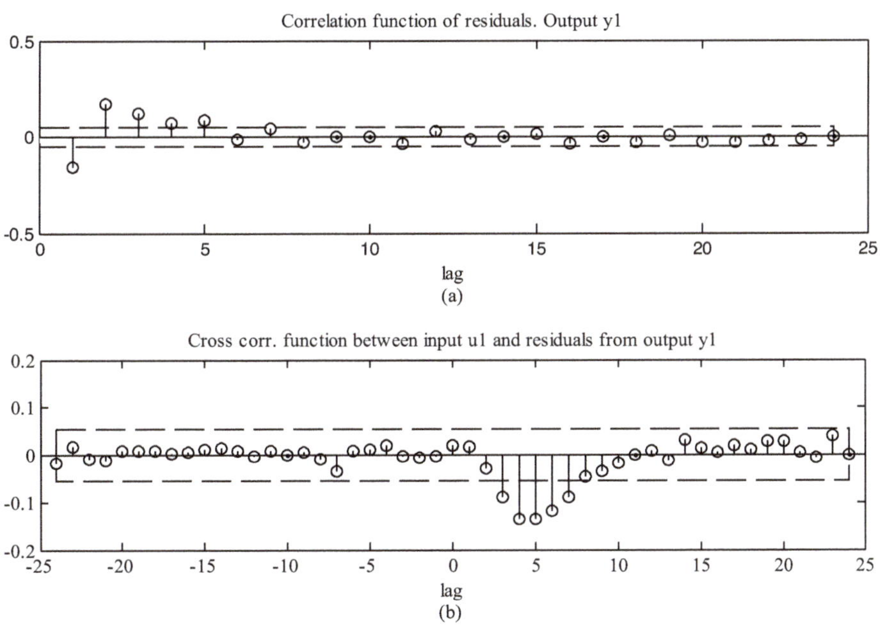

Fig. 5.28 Correlation test for model arx1

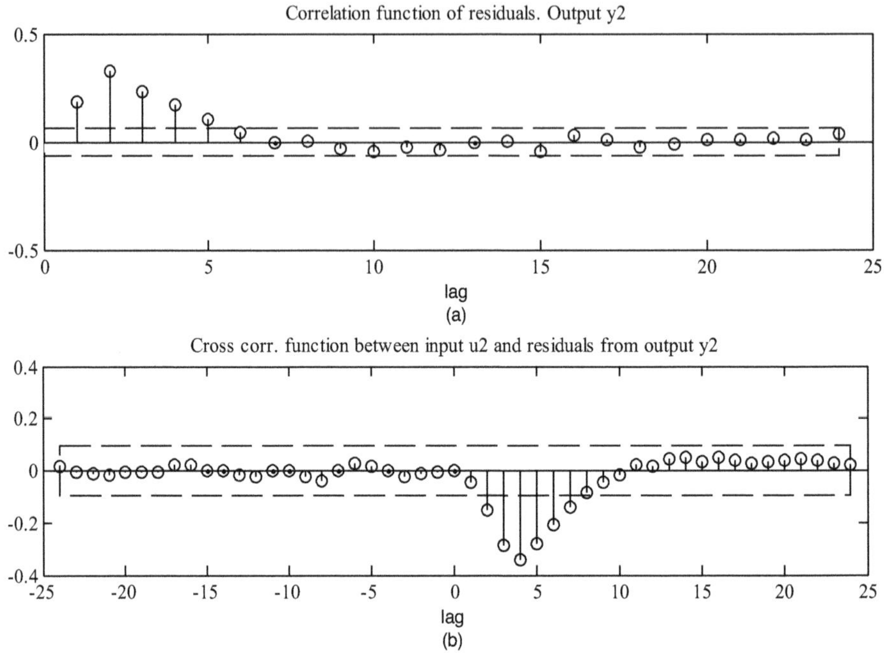

Fig. 5.29 Correlation test for model arx2

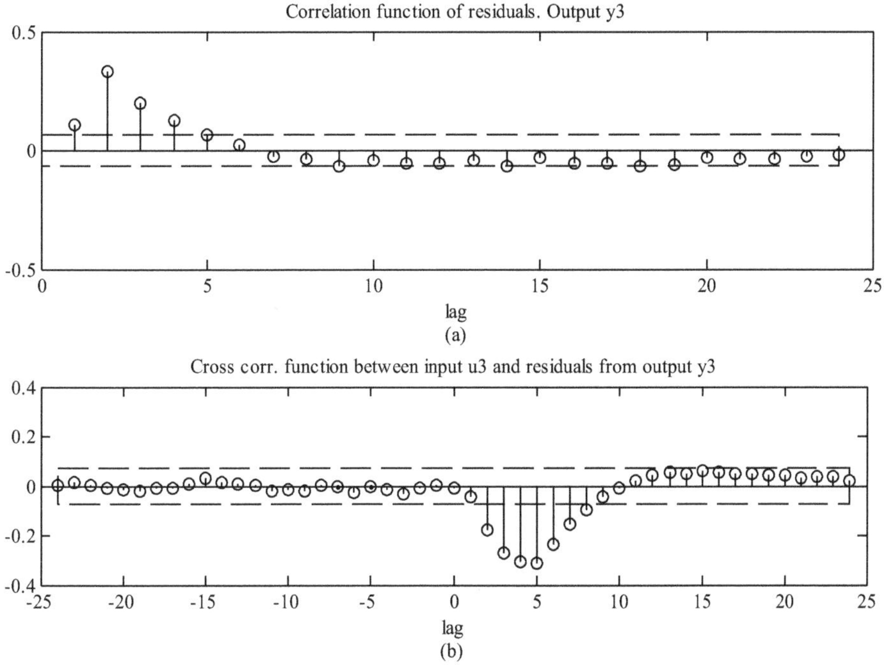

Fig. 5.30 Correlation test for model arx3

Table 5.8 Summary of correlation tests

Model's name	ACF test	CCF test
arx1	Lag 1, 2, 3, 4, 5 (5)	Lag 3, 4, 5, 6, 7 (5)
arx2	Lag 1, 2, 3, 4, 5 (5)	Lag 2, 3, 4, 5, 6, 7 ()
arx3	Lag 1, 2, 3, 4 (4)	Lag 2, 3, 4, 5, 6, 7, 8 (7)

Values in the bracketed are the total number of correlation coefficients lied outside confidence limit

Reference

1. Yusoff, Z.M.: Self-tuning Fuzzy PID Control of Hydro-diffusion Essential Oil Extraction System. Universiti Teknologi MARA, Malaysia (2014)

Chapter 6
Comparative Simulation of PID, Hybrid Fuzzy-PID, and Self-tuning Fuzzy-PID Controllers

Abstract This chapter presents a comparative simulation study of Proportional–Integral–Derivative (PID), Hybrid Fuzzy-PID (HFPID), and Self-tuning Fuzzy-PID (STFPID) controllers for steam temperature regulation in a hydro-diffusion steam distillation system with long time delay characteristics. Controller performance was evaluated using rise time, settling time, percentage overshoot, root mean square error (RMSE), set-point tracking, and load disturbance rejection. Results show that although the conventional PID controller achieves zero overshoot, it exhibits slower response and higher steady-state error compared to fuzzy-based approaches. The HFPID controller demonstrates performance improvement as the number of membership functions increases. Among all configurations, the STFPID with five membership functions achieves the best overall performance, producing the lowest RMSE, fastest settling time, and quickest recovery from load disturbances. Robustness analysis confirms that fuzzy-enhanced controllers provide superior adaptability under varying operating conditions. The findings highlight the effectiveness of integrating intelligent self-tuning fuzzy mechanisms to enhance precision, stability, and disturbance rejection in complex thermal control systems, offering a practical solution for advanced industrial temperature regulation.

6.1 Introduction

This section will be focused on the findings of the controller simulation performance that were developing to verify the proposed controllers.

Z. Mohd Yusoff et al., *Fuzzy Logic, Smart Control, and the Future of Essential Oils*, Studies in Fuzziness and Soft Computing 442,
https://doi.org/10.1007/978-981-92-2084-7_6

6.2 Simulation of Step Response Performance for PID Controller

Figure 6.1 provides the results obtained from simulation of PID controller in steam temperature control where Fig. 6.1a is referred to the control signal, and Fig. 6.1b is the output response for PID controller [1]. The result of this study shows that the output response of this controller is satisfactory. The PID controller is capable of controlling the steam temperature at the desired. However, it takes long time to settle down. The analysis performance for simulation conventional PID as tabulated in Table 6.1 [1].

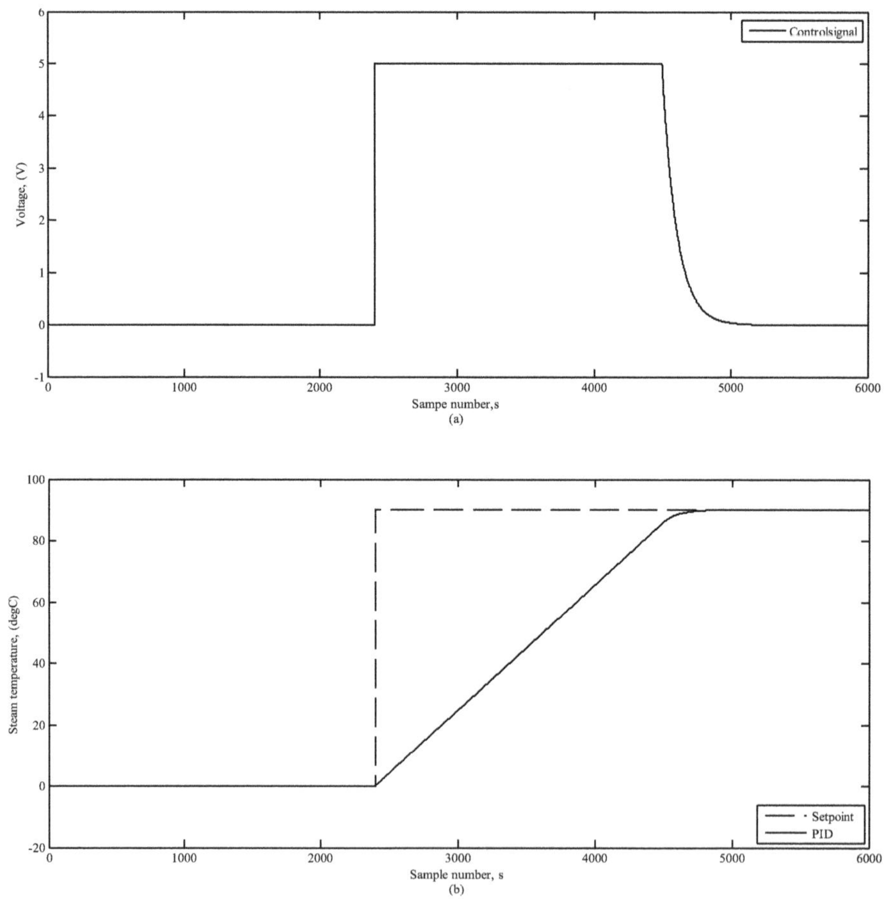

Fig. 6.1 Simulation performance of PID controller, **a** control signal, **b** output response

Table 6.1 Analysis performance for simulation of PID controller using ZN method

No.	Controller	Rise time (s)	Settling time (s)	% OS	RMSE
1	PID	1759	4587	0	0.1442

6.3 Simulation of Step Response Performance for Hybrid Fuzzy Plus PID (HFPID) Controller

Figures 6.2, 6.3, and 6.4 shows the results obtained from the simulation of hybrid fuzzy plus PID controller in steam temperature control by using 3, 5, and 7 membership functions, respectively [1]. Figure (a) is referring to the control signal, and Fig. (b) is the output response for HFPID controller. The detail performance evaluation at transient response and steady state were tabulated in Tables 6.2, 6.3, and 6.4, for HFPID-3, HFPID-5, and HDPID-7, accordingly [1]. From the results, it can be noted that all the HFPID capable of achieve the desired set point with zero percentages overshoot detected. From the simulation performances, it is apparent that the proposed HFPID controller scheme provides acceptable temperature control due to its rise time, settling time, and RMSE as compared with the conventional PID controller. However, HFPID-7 shows the best performance among all HFPID schemes in controlling steam temperature because it can control the steam temperature in a short period of rise time, lowest settling time and also has the minimum steady-state error (RMSE) throughout the whole process.

6.4 Simulation of Step Response Performance for Self-Tuning Fuzzy PID (STFPID) Controller

Figures 6.5, 6.6, and 6.7 shows the results obtained from the simulation of STFPID controller in steam temperature control by using 3, 5, and 7 membership functions, respectively. The detail performance evaluation at transient response and steady state were tabulated in Tables 6.5, 6.6, and 6.7, for STFPID-3, STFPID-5, and STFPID-7, accordingly [1].

6.5 Summary of Simulation Step Response Performances

Figure 6.8 depicts the step response simulation result for seven design controllers that were applied to control the steam temperature in hydro-diffusion steam distillation system [1]. The analysis performances for simulation of PID, HFPID-3, HFPID-5, HFPID-7, STFPID-3, STFPID-5, and STFPID-7 controllers are tabulated in Table 6.8 [1]. The results are combined together for comparison purposed. The results in the table were sorted according to the performance criteria of rise time, settling time,

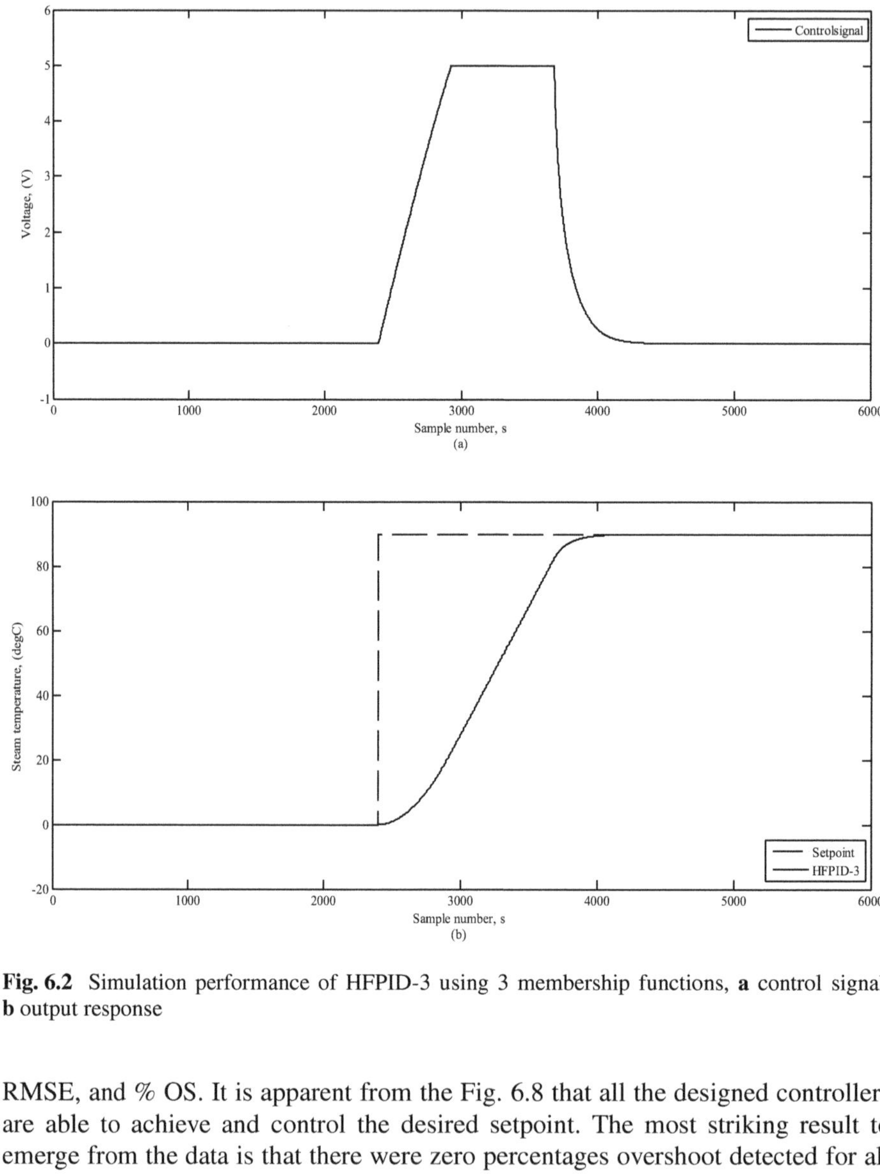

Fig. 6.2 Simulation performance of HFPID-3 using 3 membership functions, **a** control signal, **b** output response

RMSE, and % OS. It is apparent from the Fig. 6.8 that all the designed controllers are able to achieve and control the desired setpoint. The most striking result to emerge from the data is that there were zero percentages overshoot detected for all the controllers [1].

The graphical comparison between all the proposed controllers based on the rise time criteria is shown in Fig. 6.9 [1]. From the rise time result, it is clearly shown that the PID controller takes the longest time among all to achieve the desired set point which is 1759 s. The HFPID and STFPID show the better results as compared with the PID controller. The HFPID and STPFID managed to reduce half rise time significantly compared to the PID. Overall, the average range of rise time taken for controller based on fuzzy logic is around 900–930 s. These results are very

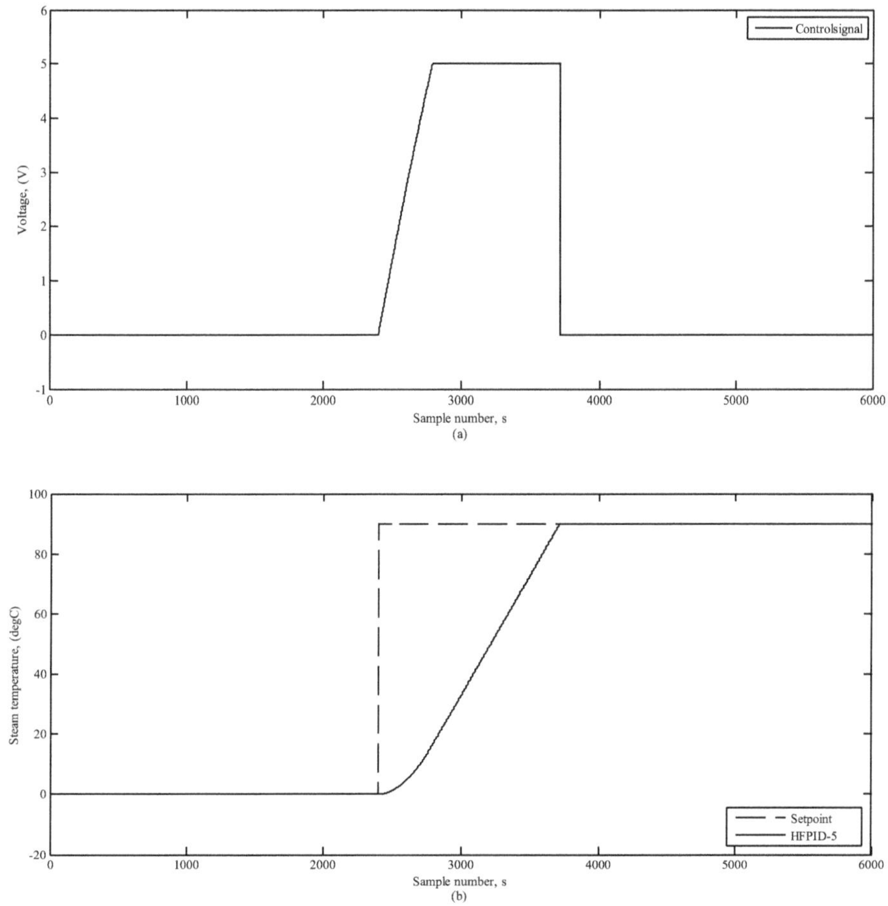

Fig. 6.3 Simulation performance of HFPID using 5 membership functions, **a** control signal, **b** output response

encouraging and can be categorized as a faster response for a system with long time delay.

Figure 6.10 graphically compares the settling time between PID, HFPID-3, HFPID-5, HFPID-7, STFPID-3, STFPID-5, and STFPID-7, consecutively [1]. Settling time is one important part in controller performance. From Fig. 6.10, it can be clearly seen that HFPID and STFPID controllers performed encouraging settling time as compared with the PID controller. PID controller took the longest settling time among all controllers' structure which is 4587 s.

Generally, for HFPID controllers, the results of this study show that significant improvement of settling time by increasing the number of membership function. The settling time is reduced with an increment of membership function. The findings are

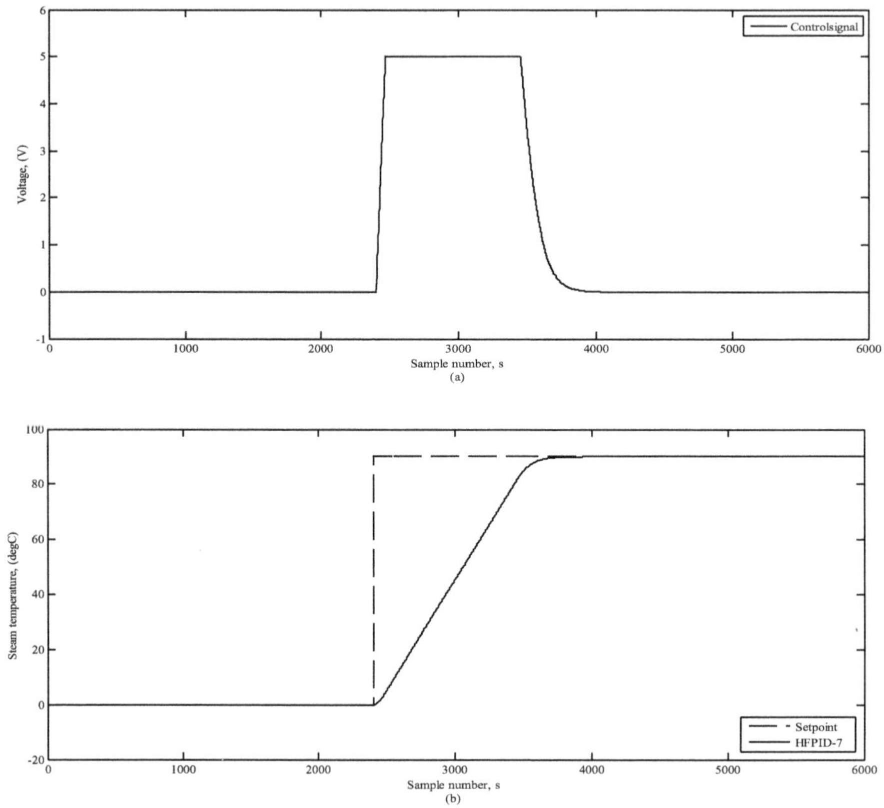

Fig. 6.4 Simulation performance of HFPID using 7 membership functions, **a** control signal, **b** output response

Table 6.2 Analysis performance for simulation of HFPID using 3 membership functions

No.	Controller	Rise time (s)	Settling time (s)	% OS	RMSE
1	HFPID-3	932	3847	0	0.0480

Table 6.3 Analysis performance for simulation of hybrid fuzzy plus PID using 5 membership functions

No.	Controller	Rise time (s)	Settling time (s)	% OS	RMSE
1	HFPID-5	912	3692	0	0.0224

Table 6.4 Analysis performance for simulation of HFPID using 7 membership functions

No.	Controller	Rise time (s)	Settling time (s)	% OS	RMSE
1	HFPID-7	900	3592	0	0.0200

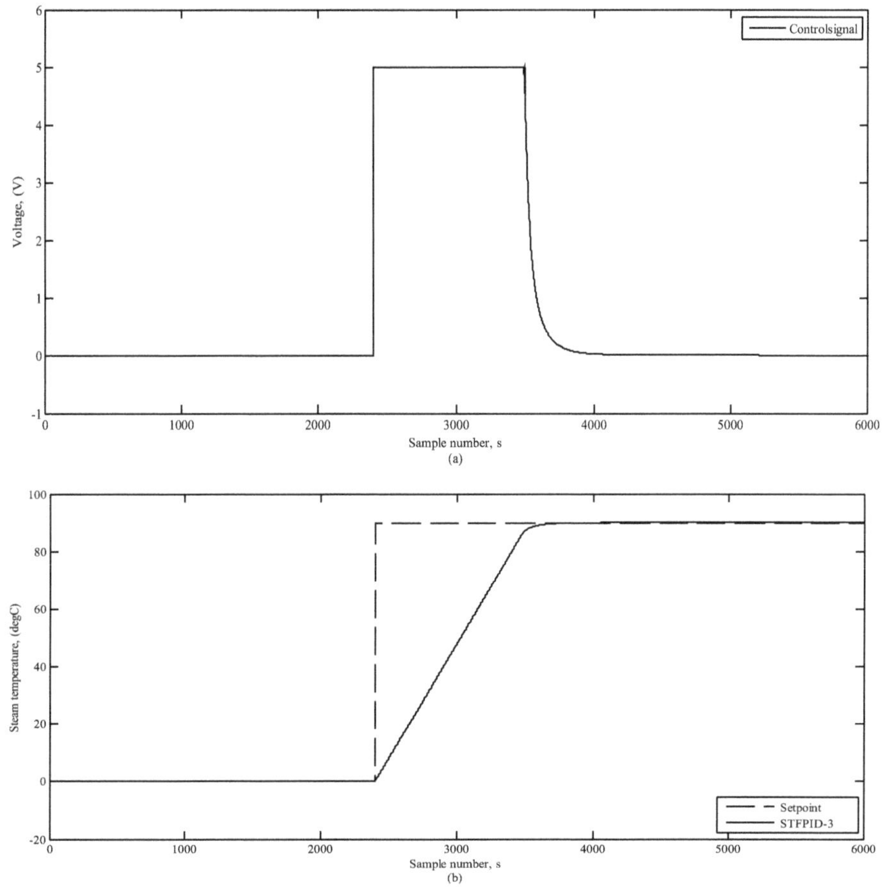

Fig. 6.5 Simulation performance of STFPID using 3 membership functions, **a** control signal, **b** output response

consistent with the rise time result for HFPID controller with HFPID-3, HFPID-5, and HFPID-7 took 3847 s, 3692 s, and 3592 s respectively.

On the other hand, for STFPID controllers, the numbers of membership function don't influence the settling time performance. It's contrary to expectations. Overall, STFPID-5 is outperformed as compared with all designed controllers by took 3504 s to settle down. It is obviously observed that the difference between PID and STFPID-5 controller was statistically significant. It can also be seen that the settling time for STFPID-3, and STFPID-7 is slightly higher than STFPID-5 with 3539 s and 3574 s, respectively. All STFPID controllers are outperformed in terms of settling time criteria as compared with the PID and HFPID schemes.

The graphical comparison of RMSE result for simulation performances between PID, HFPID-3, HFPID-5, HFPID-7, STFPID-3, STFPID-5, and STFPID-7 is illustrated in Fig. 6.11 [1]. Analysing the RMSE results, all controllers produce a good

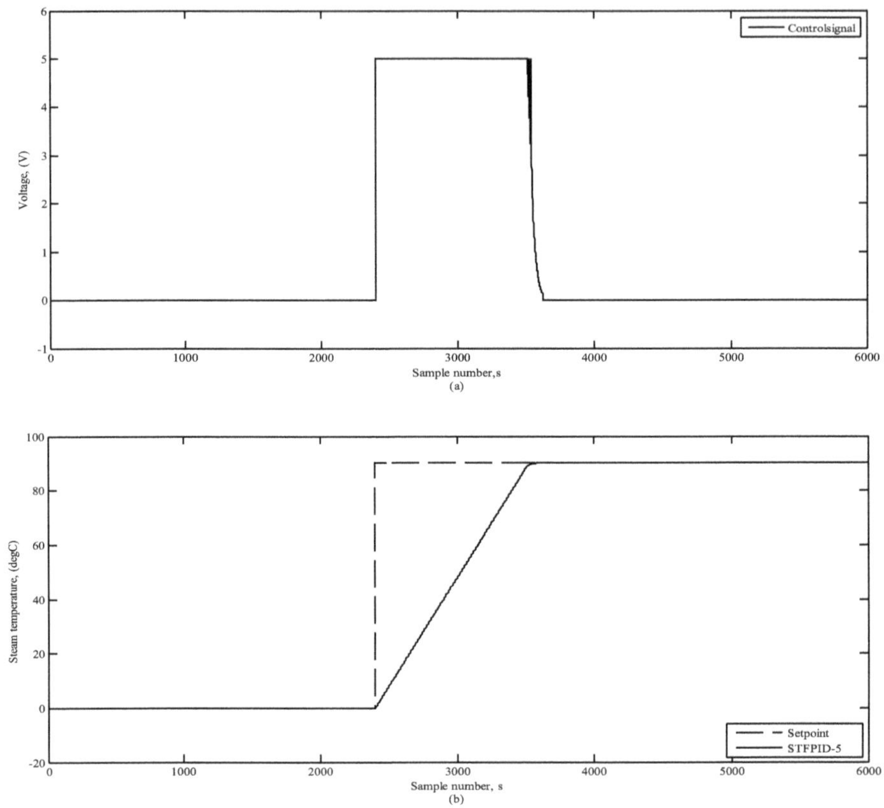

Fig. 6.6 Simulation performance of STFPID using 5 membership functions, **a** control signal, **b** output response

result since the RMSE value for all controllers is below 1, which consider as small. However, the RMSE results depict in Fig. 6.11 generally agrees with the rise time and settling time results where the PID controller is slightly higher than other controllers. The RMSE for PID controller is 0.1442.

Referring the RMSE values for HFPID controller schemes, show preferences on the HFPID-7 with 0.02, followed by HFPID-5 with 0.0224 and HFPID-3 with 0.048. This is as expected where increase the membership function will improve the RMSE result. Overall, the RMSE results show that the STFPID-5 has advantages over all controllers since it produces the smallest value of RMSE, which is 0.0141. The discrepancies between all controllers are very significant. It shows that the response at a steady state is very precise. Thus, the magnitude of temperature fluctuation at a steady state is significantly smaller. On the other hand, the RMSE value for STFPID-3 and STFPID-7 are 0.0575 and 0.0529, respectively.

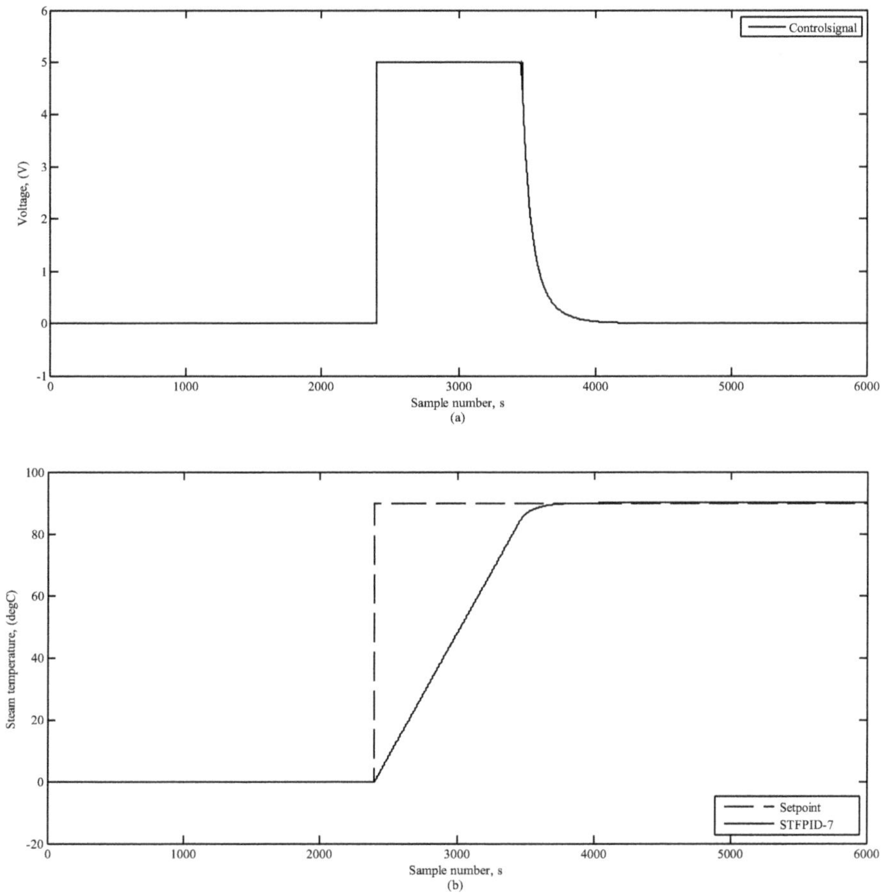

Fig. 6.7 Simulation performance of STFPID using 7 membership functions, **a** control signal, **b** output response

Table 6.5 Analysis performance for simulation of STFPID using 3 membership functions

No.	Controller	Rise time (s)	Settling time (s)	% OS	RMSE
1	STFPID-3	900	3539	0	0.0575

Table 6.6 Analysis performance for simulation of STFPID using 5 membership functions

No.	Controller	Rise time (s)	Settling time (s)	% OS	RMSE
1	STFPID-5	900	3504	0	0.0141

Table 6.7 Analysis performance for simulation of STFPID controller using 7 membership functions

No.	Controller	Rise time (s)	Settling time (s)	% OS	RMSE
1	STFPID-7	900	3574	0	0.0529

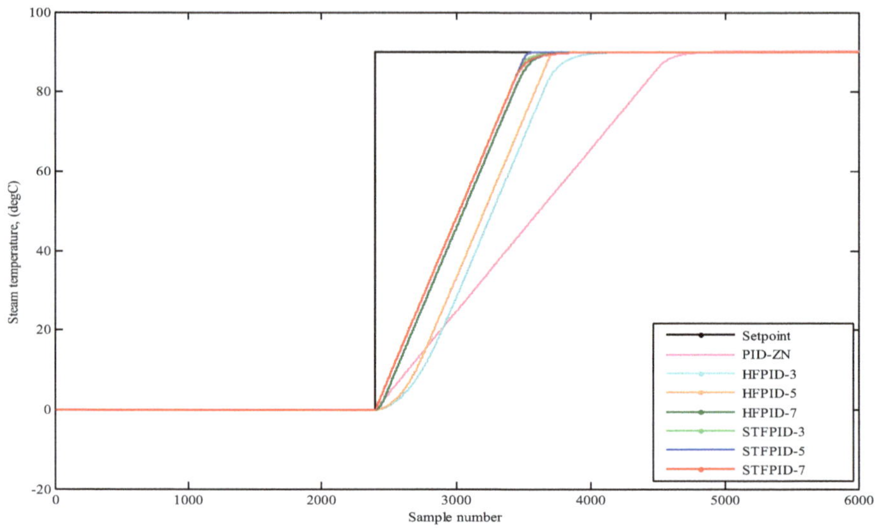

Fig. 6.8 Simulation performances of PID, HFPID-3, HFPID-5, HFPID-7, STFPID-3, STFPID-5, and STFPID-7 controller for steam temperature control

Table 6.8 Analysis performance for simulation of PID, HFPID-3, HFPID-5, HFPID-7, STFPID-3, STFPID-5, and STFPID-7 controller

No.	Controller	Rise time (s)	Settling time (s)	RMSE	% OS
1	PID	1759	4587	0.1442	0
2	HFPID-3	932	3847	0.0480	0
3	HFPID-5	912	3692	0.0224	0
4	HFPID-7	900	3592	0.0200	0
5	STFPID-3	900	3539	0.0575	0
6	*STFPID-5*	*900*	*3504*	*0.0141*	*0*
7	STFPID-7	900	3574	0.0529	0

The italic values are the best result between all controllers

6.6 Robustness Test for Simulation

This section will be discussed on the result of the stability and robustness performance for each proposed controller. Two robustness tests were carried out, which are set point tracking and introduce load disturbance.

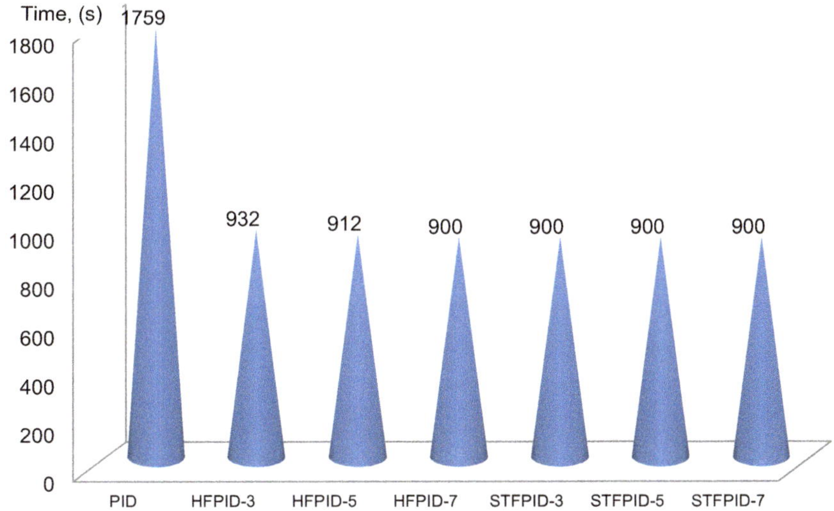

Fig. 6.9 Rise time of simulation implementation for all controllers

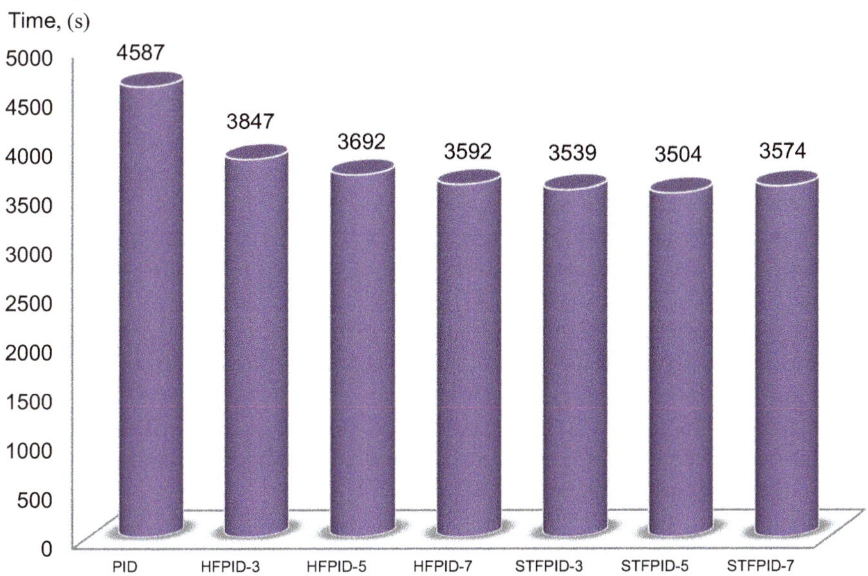

Fig. 6.10 Settling time of simulation implementation for all controllers

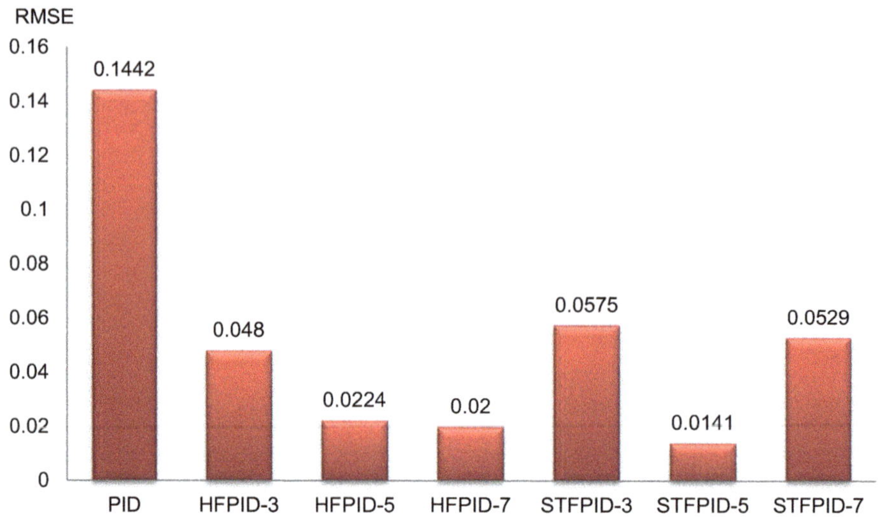

Fig. 6.11 RMSE of simulation implementation for all controllers

6.6.1 Set Point Tracking

This study is to analyze the ability of conventional PID controller and to evaluate the degree of improvement given by HFPID and STFPID controller as compared to conventional PID controller in tracking the different set point during running process. Three desired set point were set, which are 60 °C, 80 °C, and 90 °C, respectively. Each set point setting provides a different characteristic. Each proposed controller were analyzed based on rise time, settling time, %OS, and RMSE for each desired set point.

Figure 6.12 depicted the result for simulation performance of PID controller on set point tracking for steam temperature control where Fig. 6.12a is the control signal, and Fig. 6.12b is the output response [1]. The analysis performance for PID controller was tabulated in Table 6.9 [1]. From the data, it is apparent that the PID controller can reach each desired set point with no different of overshoot detected but produce slightly significant discrepancies of RMSE.

Figures 6.13, 6.14, and 6.15 depicts the output response of HFPID controller using 3, 5, and 7 membership functions on a set point tracking test, consecutively [1]. Figure (a) refers to the control signal, and Fig. (b) refers to the output response for respective controllers. Meanwhile, the analysis performances were summarized in Tables 6.10, 6.11, and 6.12, respectively [1]. The results in the figures illustrated that the HFPID-3; HFPID-5, and HFPID-7 controller produces a good results and able to sustain their performances in tracking the changes of operating point. However, further analysis showed that the HFPID-5 give better tracking set point by provide fastest rise time, shorten settling time and smallest RMSE values, followed by HFPID-7 and HFPID-3. Analyzing the %OS, all HFPID controllers recorded zero percentages overshoot for

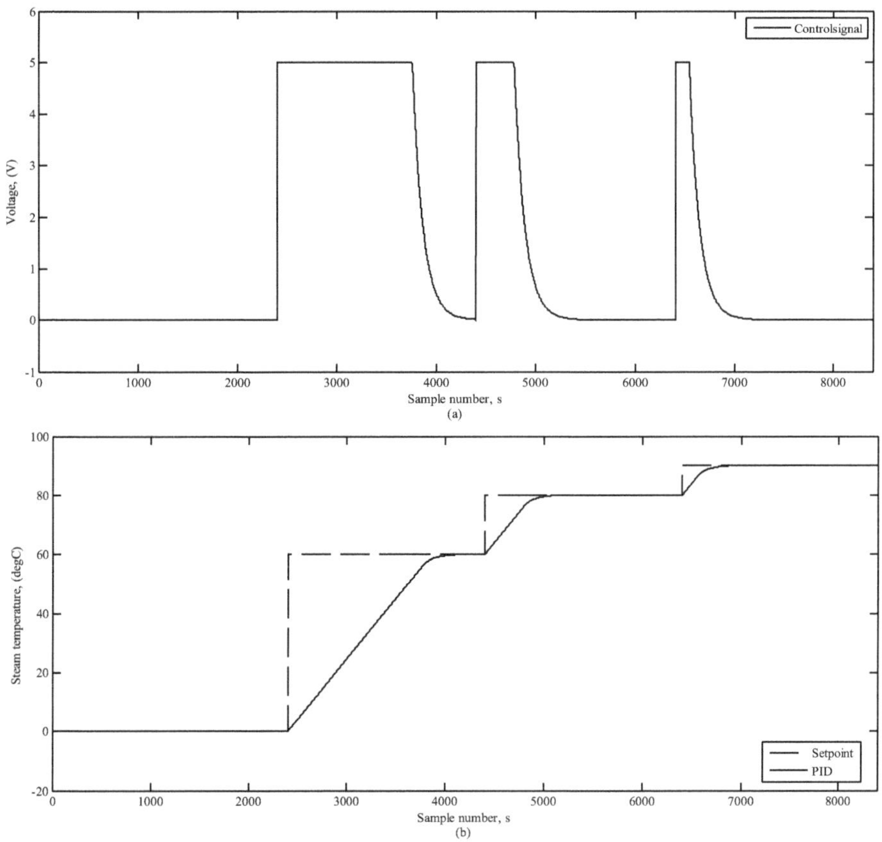

Fig. 6.12 Simulation performance for PID controller on set point tracking

Table 6.9 Analysis performance for simulation of PID controller on set point tracking

No.	Temperature (°C)	Rise time (s)	Settling time (s)	% OS	RMSE
1	60	1171	1498	0	0.2927
2	80	416	639	0	0.2147
3	90	270	468	0	0.2182

each set point setting. Currently, HFPID-3 and HFPID-7 show comparable performances of RMSE values. The results show there is only slightly significant different of RMSE between them.

Figures 6.16, 6.17, and 6.18 presents the result for simulation performance of STFPID-3, STFPID-5, and STFPID-7 controller on set point tracking for steam temperature control where Figure (a) is the control signal, and Fig. (b) is the output response [1]. Tables 6.13, 6.14, and 6.15 tabulated the analysis performances for respective controllers [1]. From the results, it can be noted that the STFPID-3,

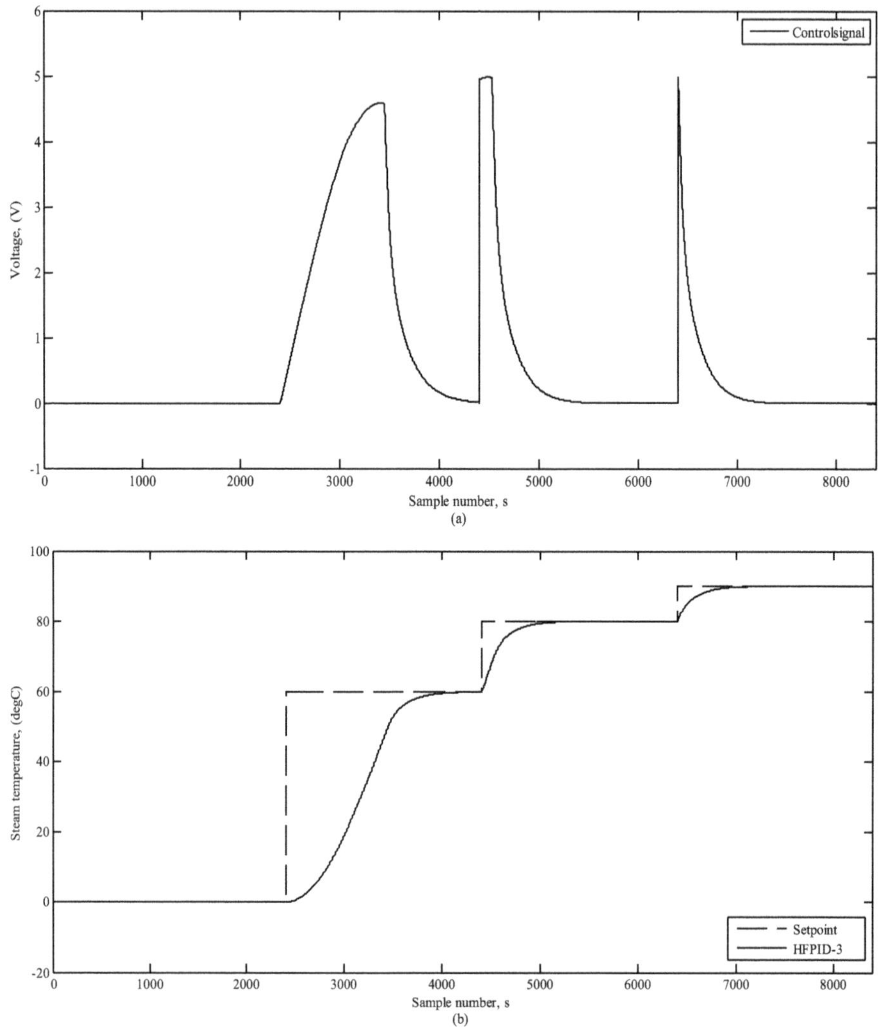

Fig. 6.13 Simulation performance for HFPID-3 on set point tracking

STFPID-5, and STFPID-7 controller produces a satisfactory results and able to adapt the changes of set point during running process. However, the STFPID-5 is found slightly better because can track the new set point either large step change or low step change with fast rise time, fast settling time and smallest steady-state error (RMSE). Analyzing the %OS, all STFPID controllers recorded zero percentages overshoot for each operating set point. Referring the RMSE results for STFPID-3 and STFPID-7, it shows there are only slightly significant discrepancies of RMSE between them.

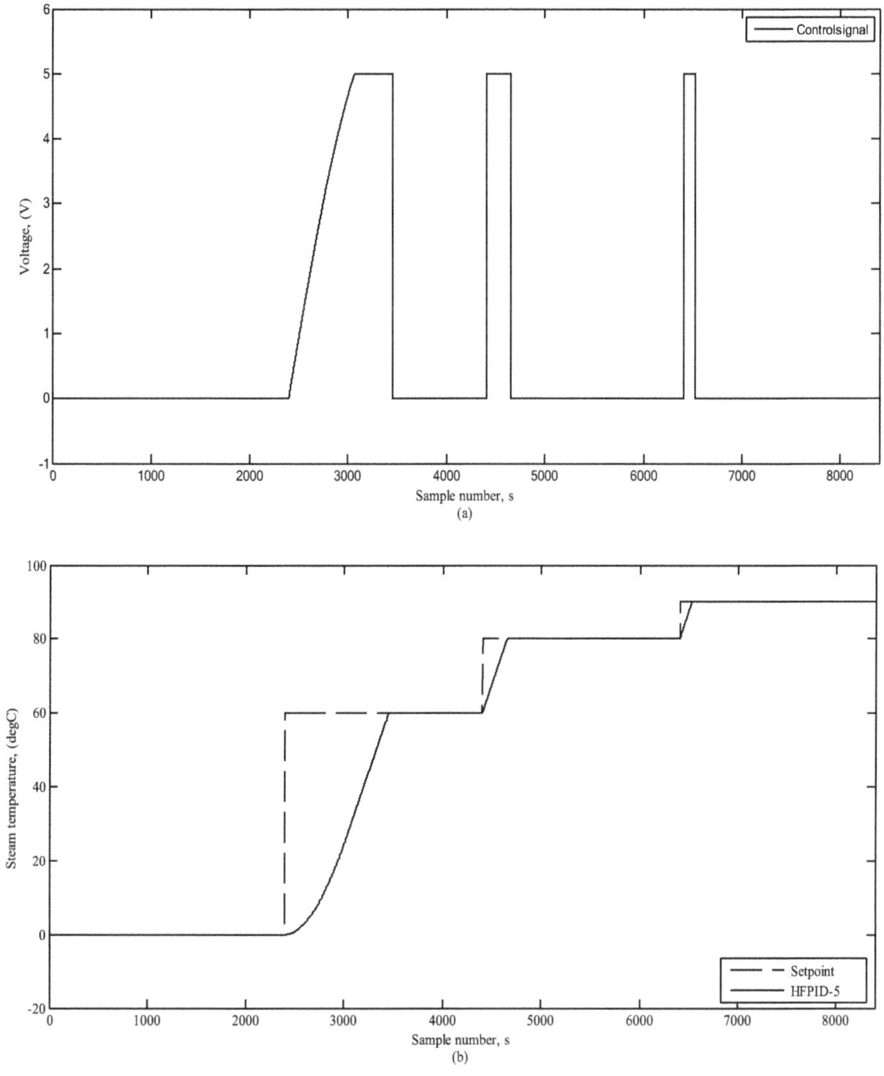

Fig. 6.14 Simulation performance for HFPID-5 on set point tracking

Figure 6.19 presents the combination result for set point tracking test for all proposed controllers [1]. Generally, the proposed PID, HFPID-3, HFPID-5, HFPID-7, STFPID-3, STFPID-5, and STFPID-7 controllers' show a satisfactory results because can track the changes in set point whether in small or large set point change. However, HFPID and STFPID offer better performances in terms of rise time, settling time, and steady-state error (RMSE).

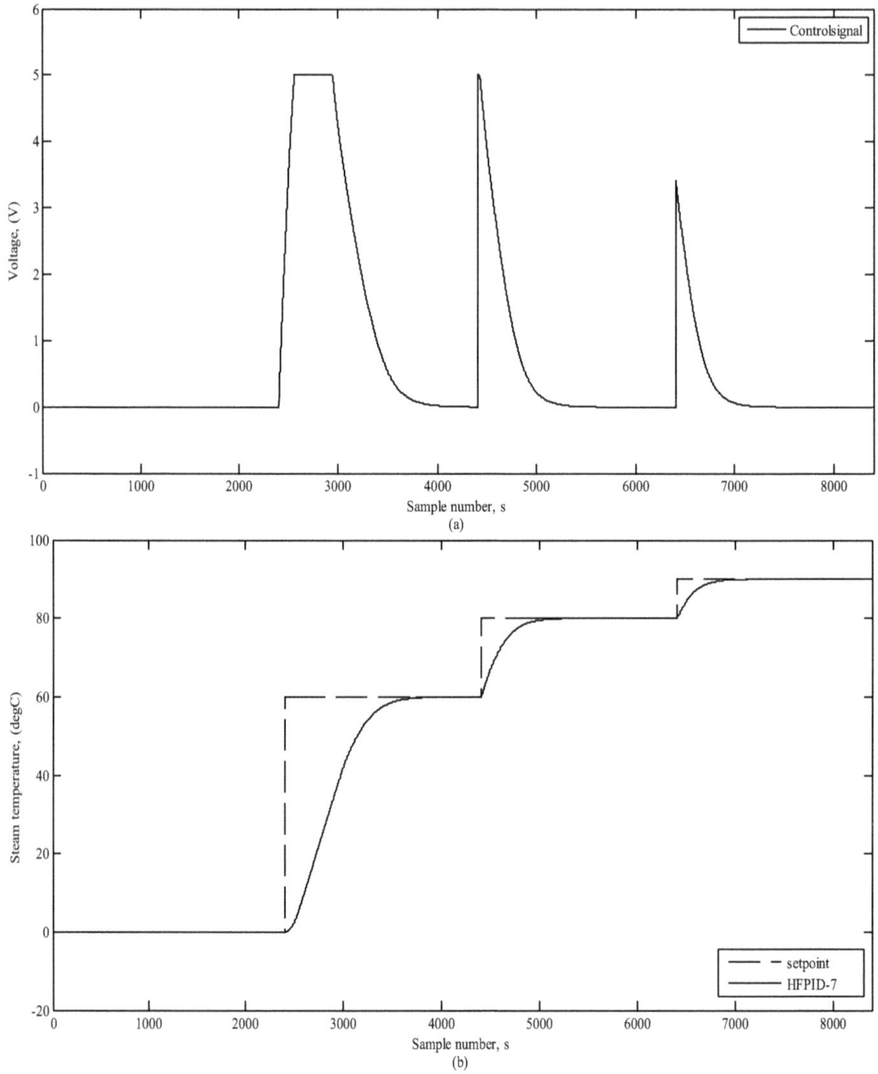

Fig. 6.15 Simulation performance for HFPID-7 on set point tracking

Table 6.10 Analysis performance for simulation of HFPID-3 on set point tracking

No.	Temperature (°C)	Rise time (s)	Settling time (s)	% OS	RMSE
1	60	799	1431	0	0.1833
2	80	386	672	0	0.0854
3	90	362	639	0	0.0480

Table 6.11 Analysis performance for simulation of HFPID-5 on set point tracking

No.	Temperature (°C)	Rise time (s)	Settling time (s)	% OS	RMSE
1	60	686	1037	0	0.0975
2	80	200	249	0	0.0624
3	90	100	127	0	0.0616

Table 6.12 Analysis performance for simulation of HFPID-7 on set point tracking

No.	Temperature (°C)	Rise time (s)	Settling time (s)	% OS	RMSE
1	60	702	1120	0	0.1224
2	80	381	631	0	0.091
3	90	319	548	0	0.073

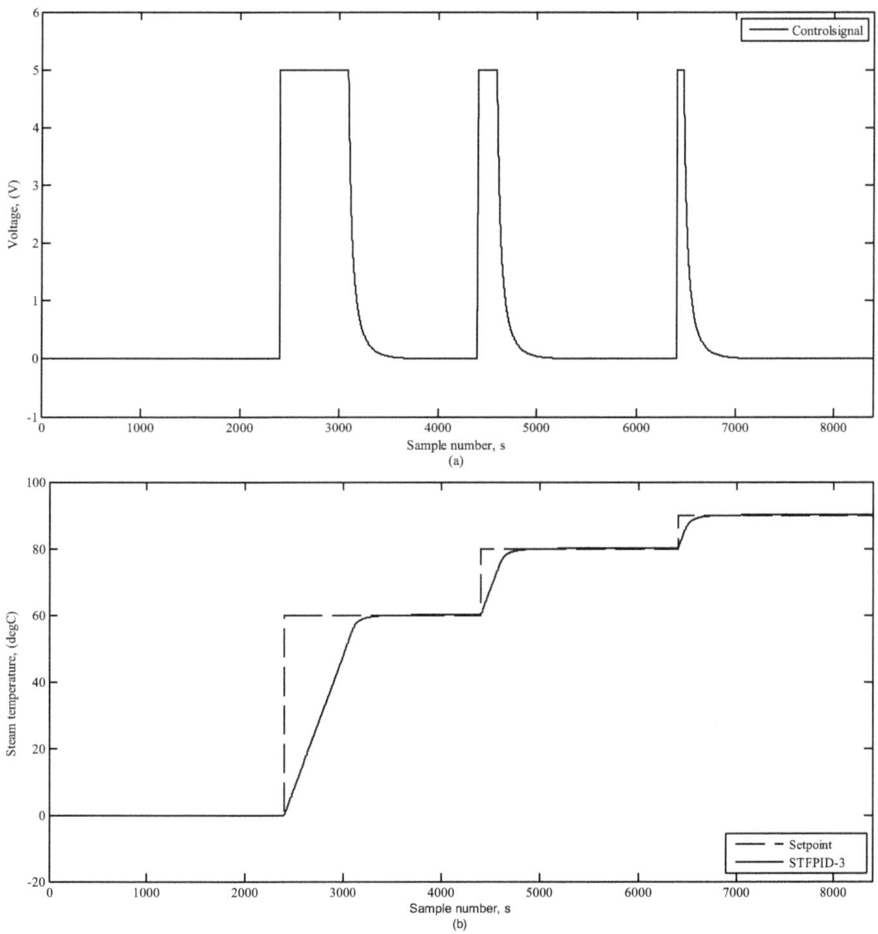

Fig. 6.16 Simulation performance for STFPID-3 on set point tracking

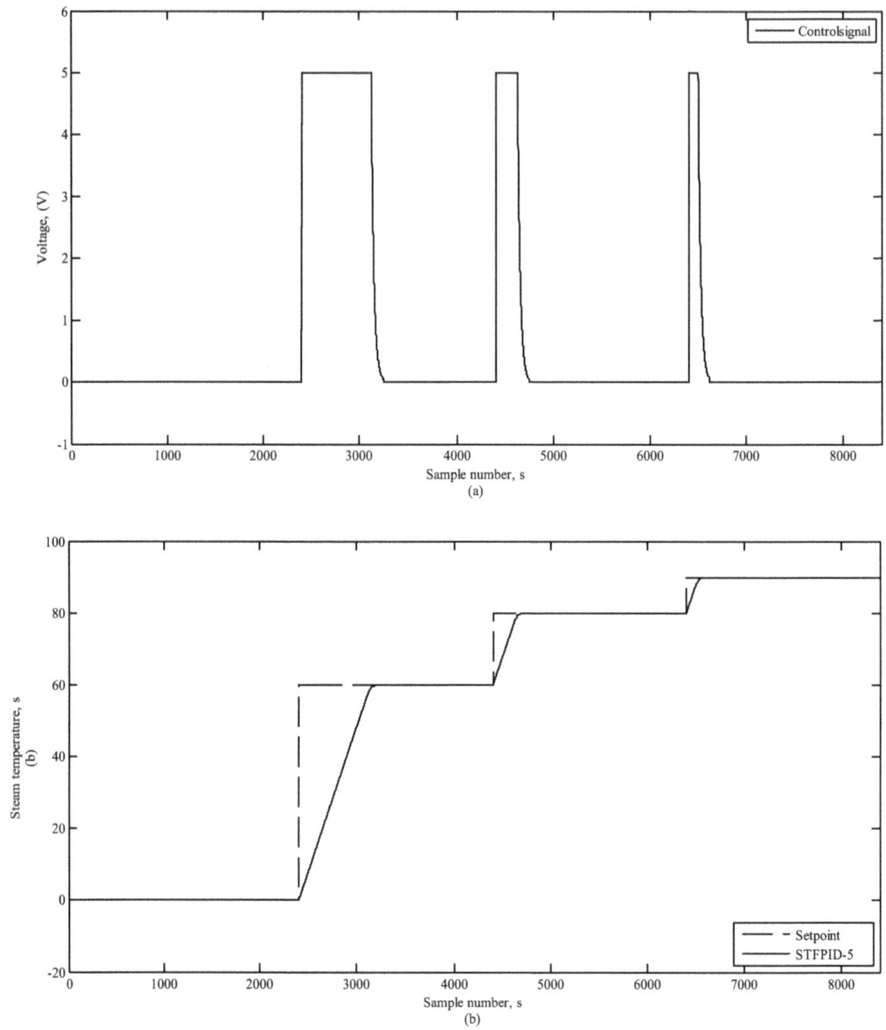

Fig. 6.17 Simulation performance for STFPID-5 on set point tracking

6.6.2 Load Disturbance

The performances of the various proposed controllers are examined by introduced disturbance. This study is to stress the effectiveness and the capability of conventional PID controller and to evaluate the extend of improvement given by HFPID controller and STFPID controller as compared to the conventional PID controller in recovering load disturbance during running process. The disturbance test was done by injecting 10 °C down from the desired set point starting at 5000th samples. Robustness of the controller design will be based on the how fast output response tackles back to the

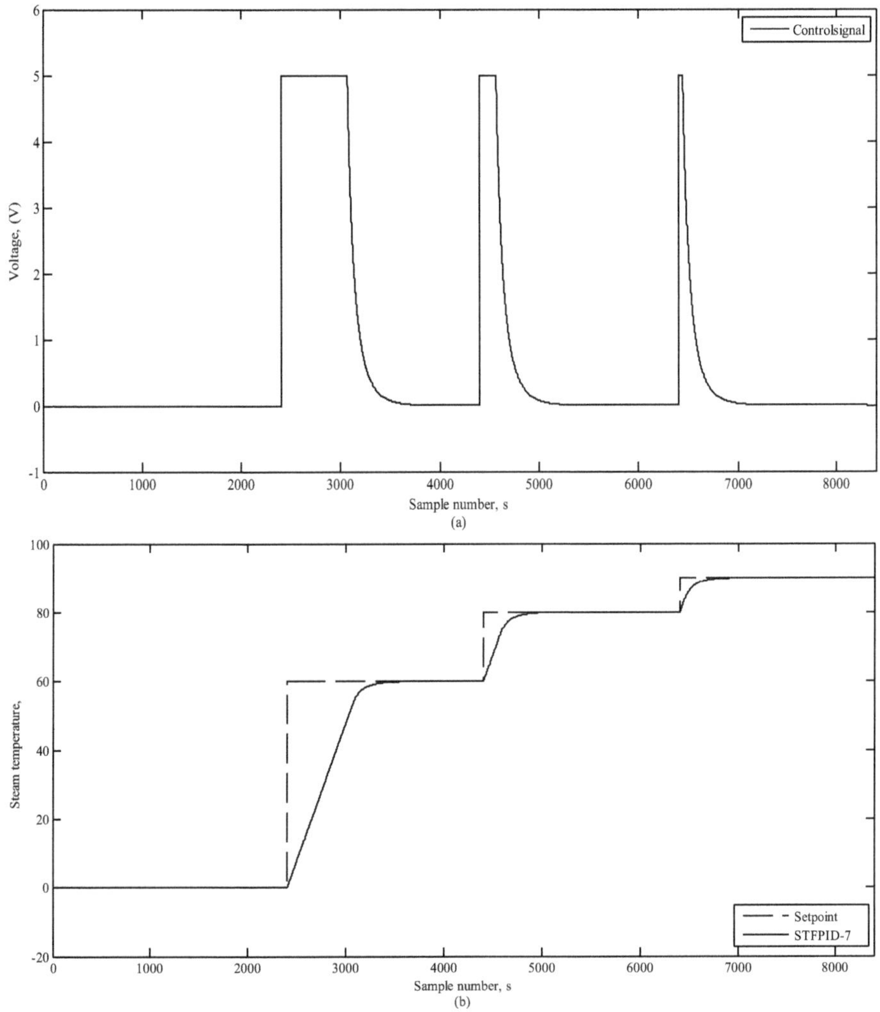

Fig. 6.18 Simulation performance for STFPID-7 on set point tracking for steam temperature control

Table 6.13 Analysis performance for simulation of STFPID-3 on set point tracking

No.	Temperature (°C)	Rise time (s)	Settling time (s)	% OS	RMSE
1	60	600	804	0	0.1000
2	80	228	425	0	0.0877
3	90	181	386	0	0.0774

Table 6.14 Analysis performance for simulation of STFPID-5 controller on set point tracking

No.	Temperature (°*C*)	Rise time (s)	Settling time (s)	% OS	RMSE
1	60	600	742	0	0.0458
2	80	200	270	0	0.0412
3	90	106	162	0	0.0387

Table 6.15 Analysis performance for simulation of STFPID-7 controller on set point tracking

No.	Temperature (°C)	Rise time (s)	Settling time (s)	% OS	RMSE
1	60	600	847	0	0.1192
2	80	260	484	0	0.0883
3	90	228	449	0	0.0800

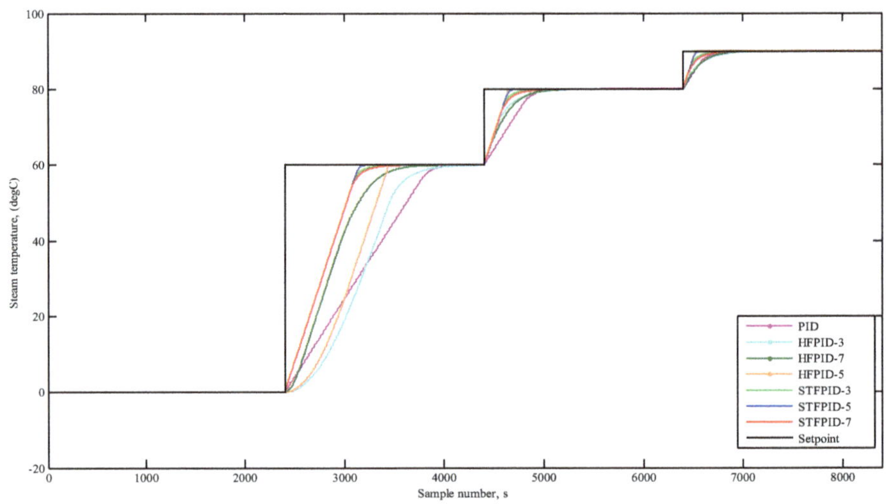

Fig. 6.19 Output response for all proposed controllers on setpoint change test

desired set point. The simulation analysis performances for PID using ZN, HFPID-3, HFPID-5, HFPID-7, STFPID-3, STFPID-5, and STFPID-7 controller were tabulated in Table 6.16 [1] (Fig. 6.20).

Figure 6.21 depicts the output response of hydro-diffusion system using HFPID-3 on recovering load disturbance [1]. The HFPID-3 controller scheme shows satisfactory result to control the steam temperature by taking 235 s to return back to the set point after injecting the disturbance.

The result of HFPID-5 on recovering load disturbance is shown in Fig. 6.22 [1]. The result demonstrates that the proposed HFPID-5 scheme take 198 s to go back to the desired set point (90 °C) when load disturbance is suddenly added at 5000th samples.

Table 6.16 Analysis performance for simulation of PID, HFPID-3, HFPID-5, HFPID-7, STFPID-3, STFPID-5, and STFPID-7 on recovering load disturbance

No.	Controller	T_{min} (°C)	Recovery time (s)
1	PID	80	270
2	HFPID-3	80	235
3	HFPID-5	80	198
4	HFPID-7	80	166
5	STFPID-3	80	182
6	STFPID-5	80	105
7	STFPID-7	80	228

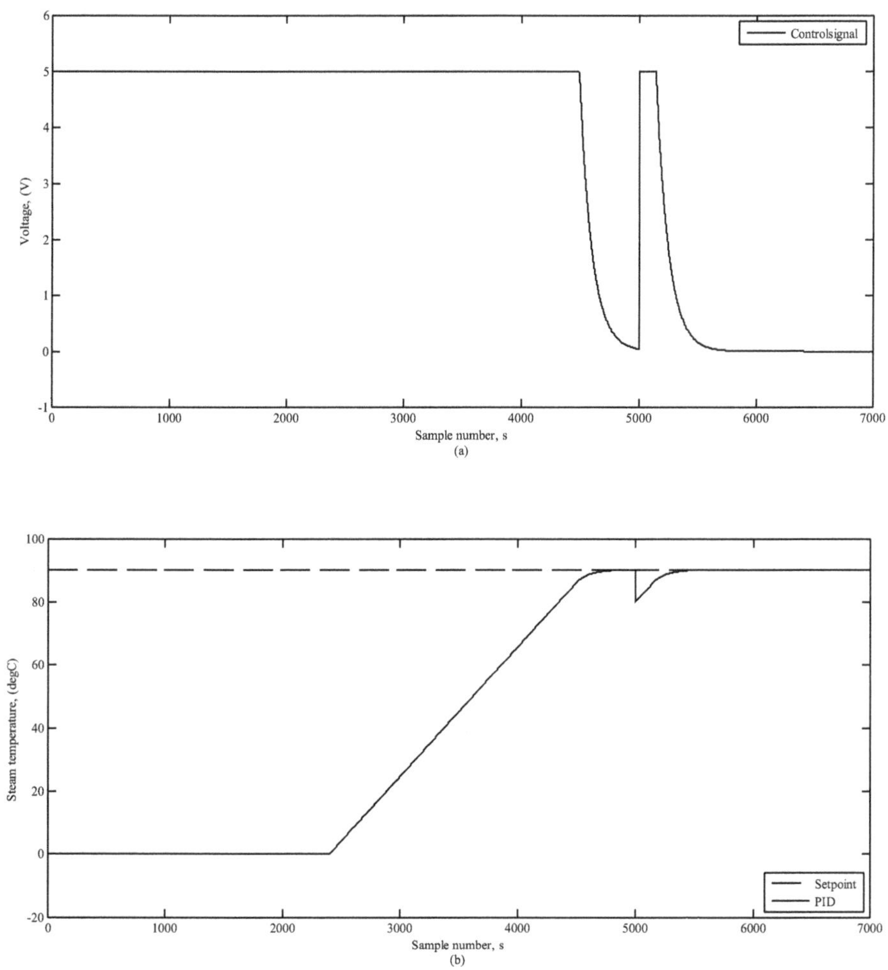

Fig. 6.20 Simulation performance of PID controller on load disturbance [1]

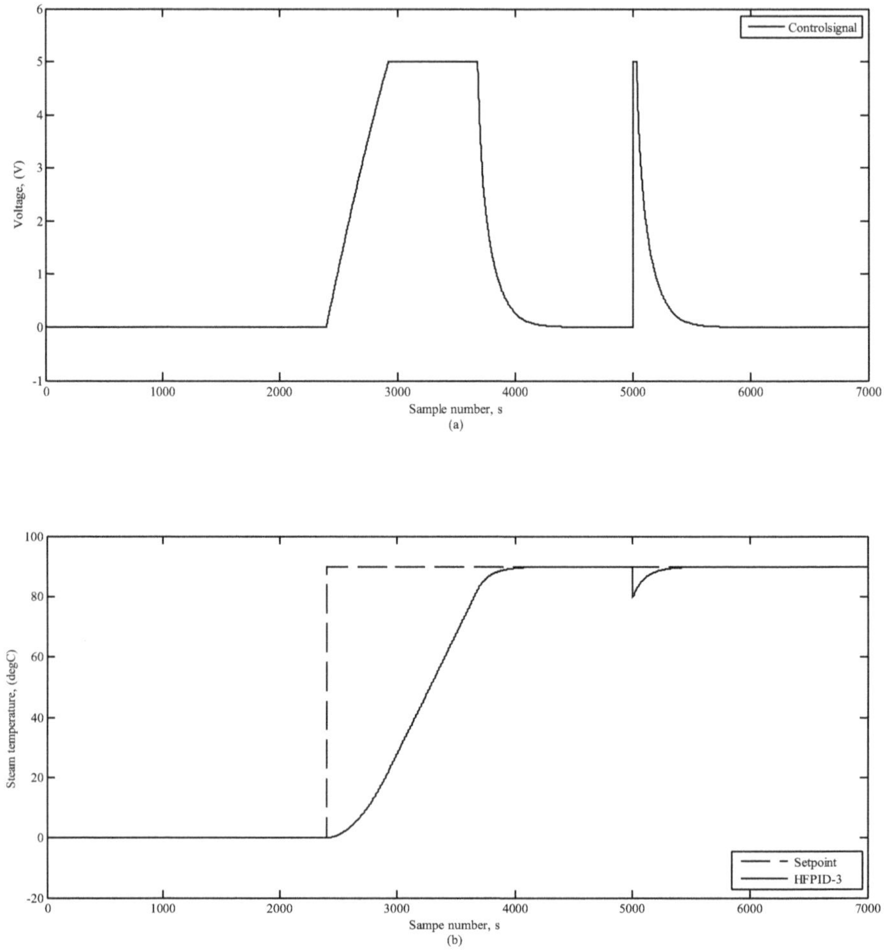

Fig. 6.21 Simulation performance of HFPID-3 on load disturbance

The result for simulation of HFPID-7 on recovering load disturbance for steam temperature control is depicted in Fig. 6.23 [1]. From the result, it can be noted that the HFPID-7 scheme able to returns to the set point by take 166 s after injecting the disturbance.

Figure 6.24 depicts the output response of hydro-diffusion system using STFPID-3 on recovering load disturbance [1]. The STFPID-3 controller scheme shows a satisfactory result to control the steam temperature by taking 182 s to return to the set point after injecting the disturbance.

The result of STFPID-5 on recovering load disturbance is shown in Fig. 6.25 [1]. From the result can be noticed that the proposed STFPID-5 scheme take 105 s to go back to the desired set point (90 °C) when load disturbance is suddenly added at 5000th samples.

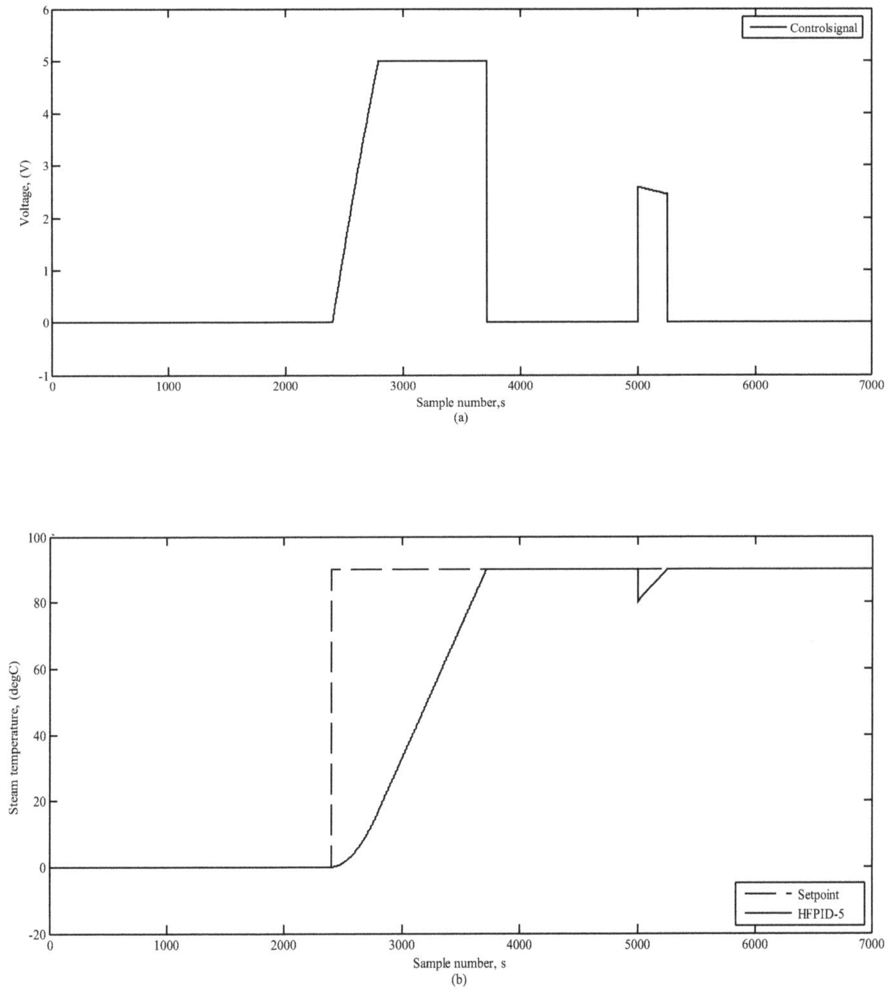

Fig. 6.22 Simulation performance of HFPID-5 on load disturbance

The result for simulation of STFPID-7 on recovering load disturbance for steam temperature control is depicted in Fig. 6.26 [1]. From the result, it can be noted that the STFPID-7 scheme able to returns to the set point by taken 228 s after injecting the disturbance.

Figure 6.27 shows the simulation results that have been combined for all developed controllers for the comparison purposed on recovering load disturbance test [1]. From the figure, it is clearly shown that before the loads are introduced, all controllers exhibit a good output response and able to meet desired set point. However, when the disturbance is suddenly added during running process, it greatly affected the output response for all controller schemes. The desired steam temperature drop and recovery

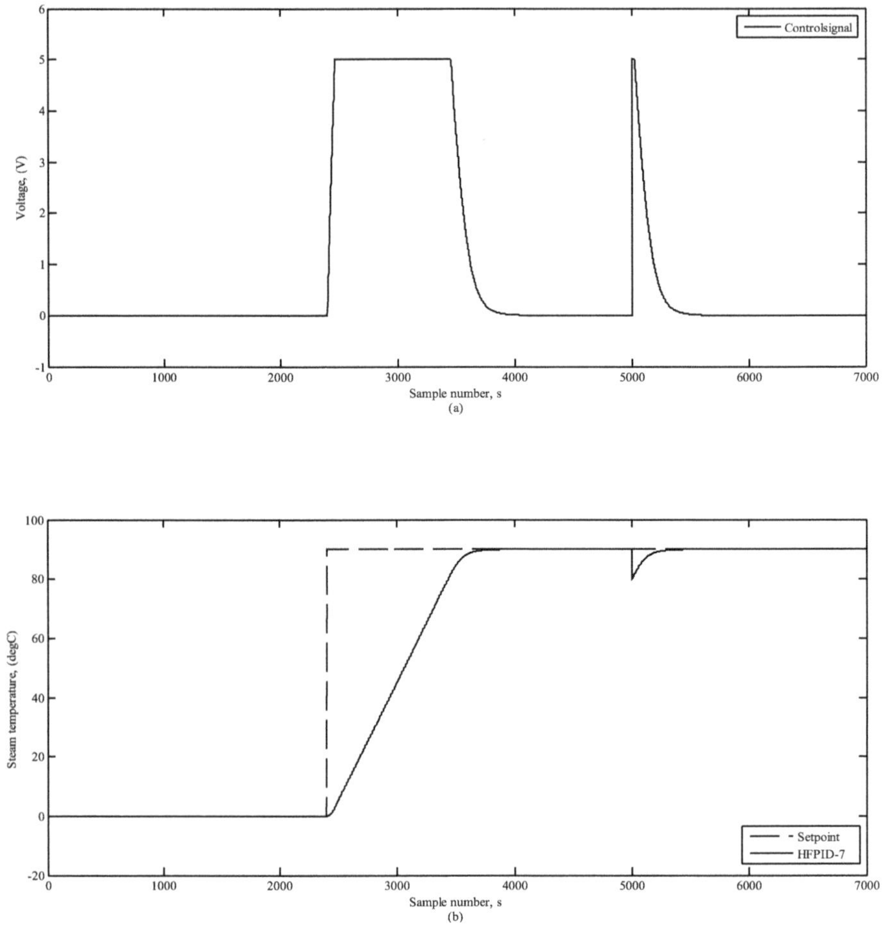

Fig. 6.23 Simulation performance of HFPID-7 on load disturbance

process appear. The detail performance's evaluation for each controller design based on time required on recovering load disturbance or rise time is tabulated in Table 6.16 and Fig. 6.28 [1].

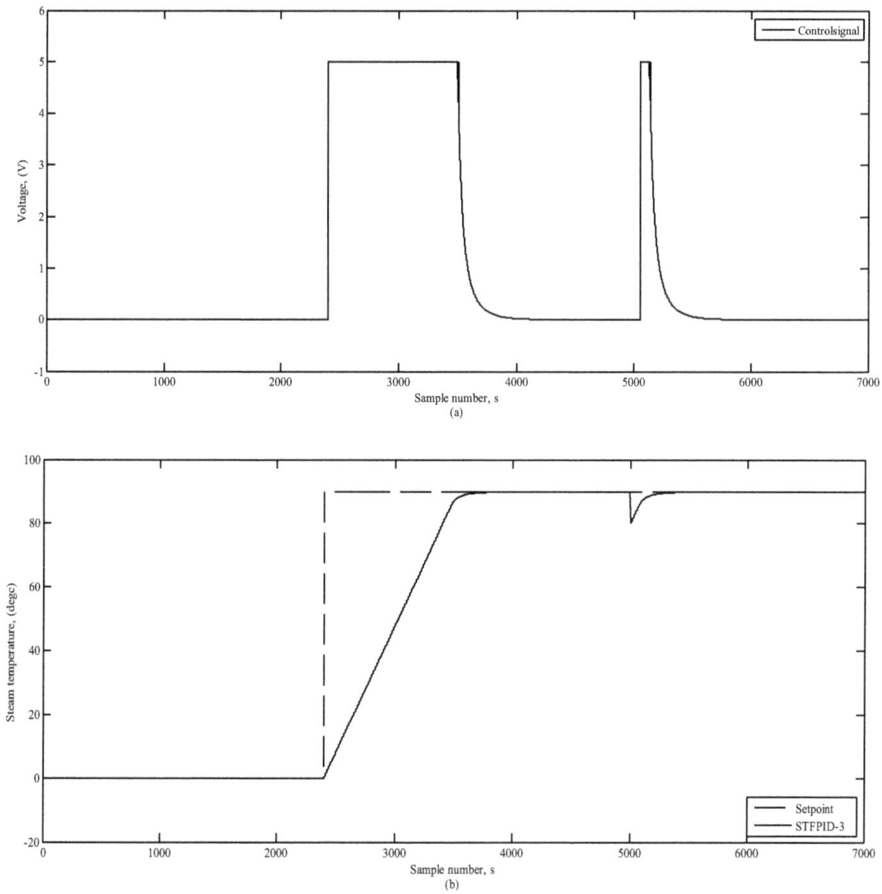

Fig. 6.24 Simulation performance of STFPID-3 on load disturbance

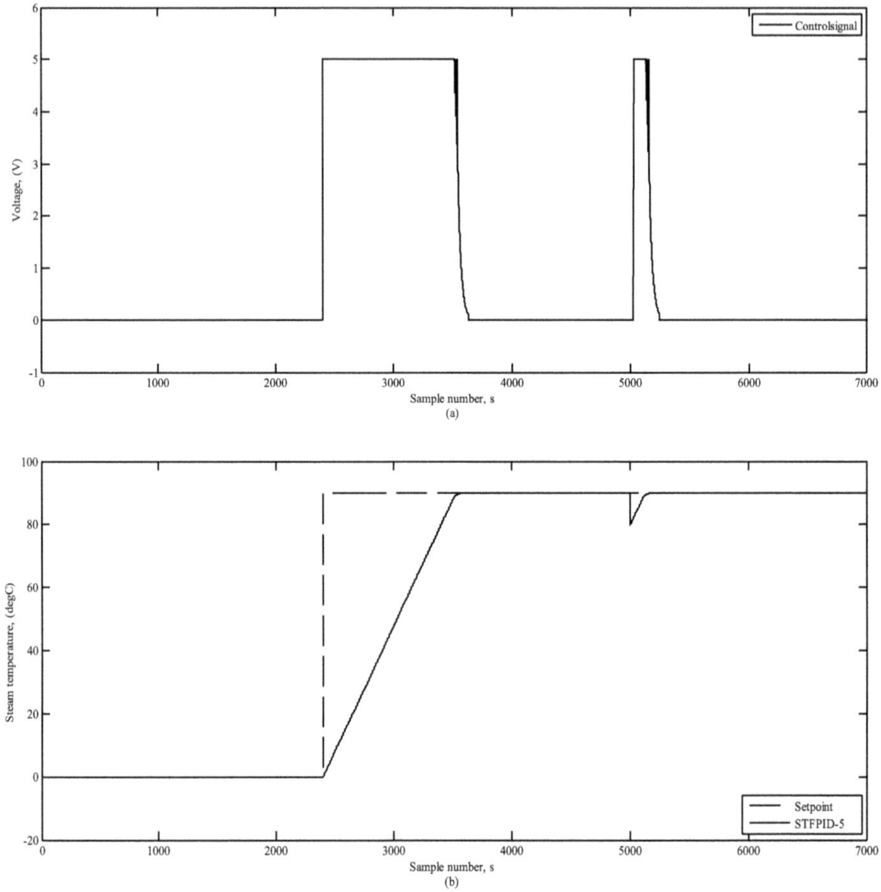

Fig. 6.25 Simulation performance of STFPID-5 on load disturbance

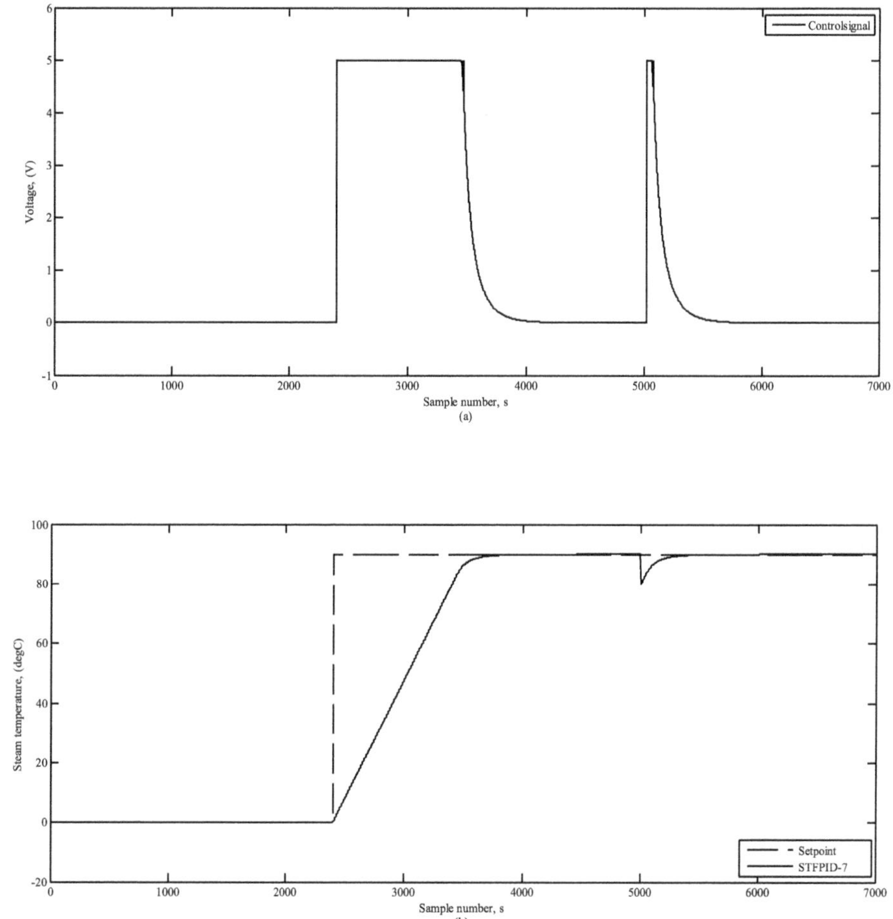

Fig. 6.26 Simulation performance of STFPID-7 on load disturbance

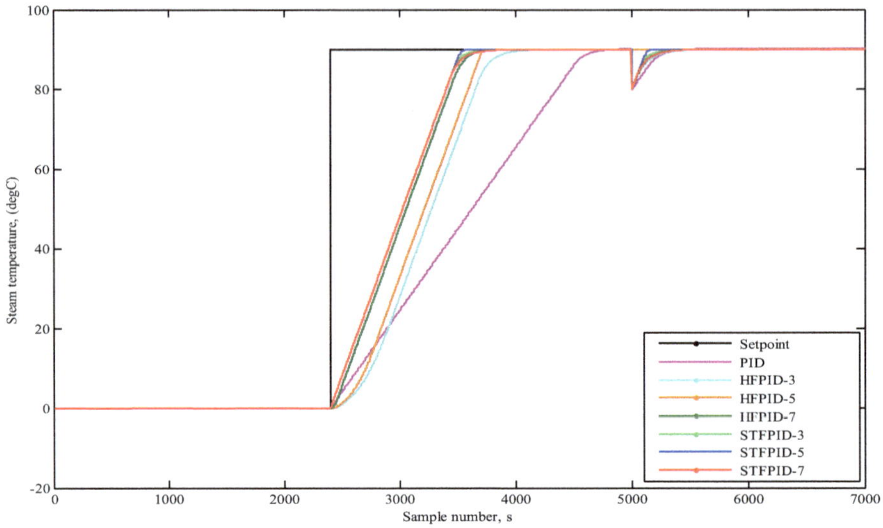

Fig. 6.27 Output response for all proposed controllers on recovering load disturbance test

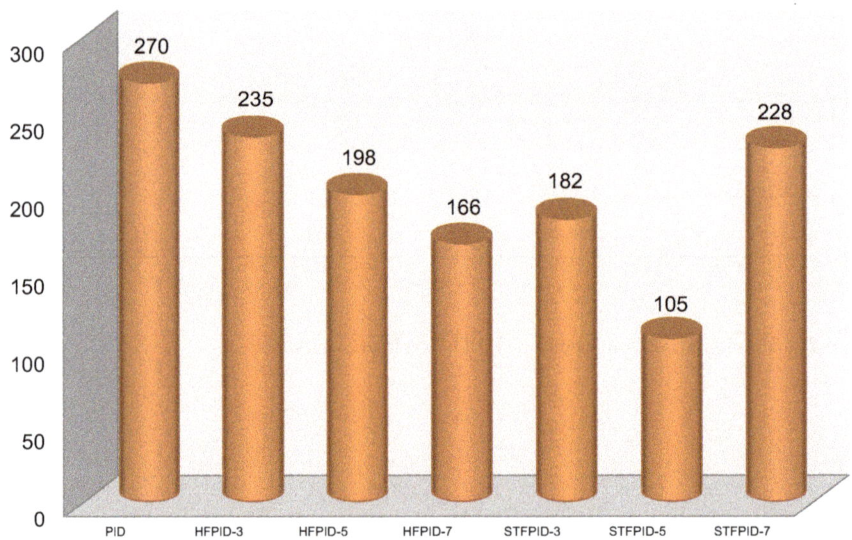

Fig. 6.28 Comparison of time required for all proposed controller on recovering load disturbance

Reference

1. Yusoff, Z.M.: Self-tuning Fuzzy PID Control of Hydro-diffusion Essential Oil Extraction System. Universiti Teknologi MARA, Malaysia (2014)

Chapter 7
Real-Time Validation of PID, Hybrid Fuzzy-PID, and Self-tuning Fuzzy-PID Controllers

Abstract Building upon the comparative simulation analysis presented in Chap. 6, this chapter provides comprehensive real-time experimental validation of PID, Hybrid Fuzzy-PID (HFPID), and Self-Tuning Fuzzy-PID (STFPID) controllers for steam temperature regulation in a hydro-diffusion steam distillation system characterized by long time delay and nonlinear dynamics. While simulation results previously demonstrated the superiority of fuzzy-enhanced strategies, practical implementation is essential to verify their industrial applicability. The controllers were deployed on an actual essential oil extraction setup and evaluated using rise time, settling time, percentage overshoot, root mean square error (RMSE), set-point tracking, and load disturbance rejection. Experimental findings confirm that fuzzy-based controllers consistently outperform the conventional PID controller under real operating conditions. HFPID-7 achieves the fastest transient response, whereas STFPID-7 produces the smallest steady-state error. In disturbance testing, HFPID-7 demonstrates remarkable robustness, reducing recovery time by more than fivefold compared to PID control. The strong agreement between simulation and real-time results validates the reliability of intelligent fuzzy integration. This study substantiates the practical significance of advanced fuzzy-PID architectures for improving stability, adaptability, and energy efficiency in complex industrial thermal processes.

7.1 Introduction

Simulation and real-time implementation are essential in process control. In this study, the performances of developed controllers are demonstrated with simulations and experiments.

Z. Mohd Yusoff et al., *Fuzzy Logic, Smart Control, and the Future of Essential Oils*, Studies in Fuzziness and Soft Computing 442,
https://doi.org/10.1007/978-981-92-2084-7_7

7.2 Real Time of Step Response for PID Controller

The finding for real-time implementation of PID controller is depicted in Fig. 7.1a the control signal and Fig. 7.1b the step response performance [1]. The result from this study shows that, the PID controller able to regulate the steam temperature in order to achieve the desired set point with large settling time and large rise time, which are 959 s and 3657 s, respectively. However, PID controller produced significant %OS, which is 0.3340. From the Fig. 7.1a, it can be observed that from the beginning of the experiment, the control signal is maintained at 5 V. However, once the output response achieved the desired set point (90 °C), the control signal start varies between 0 and 5 V. The analysis performance for real-time PID controller is summarized in Table 7.1 [1].

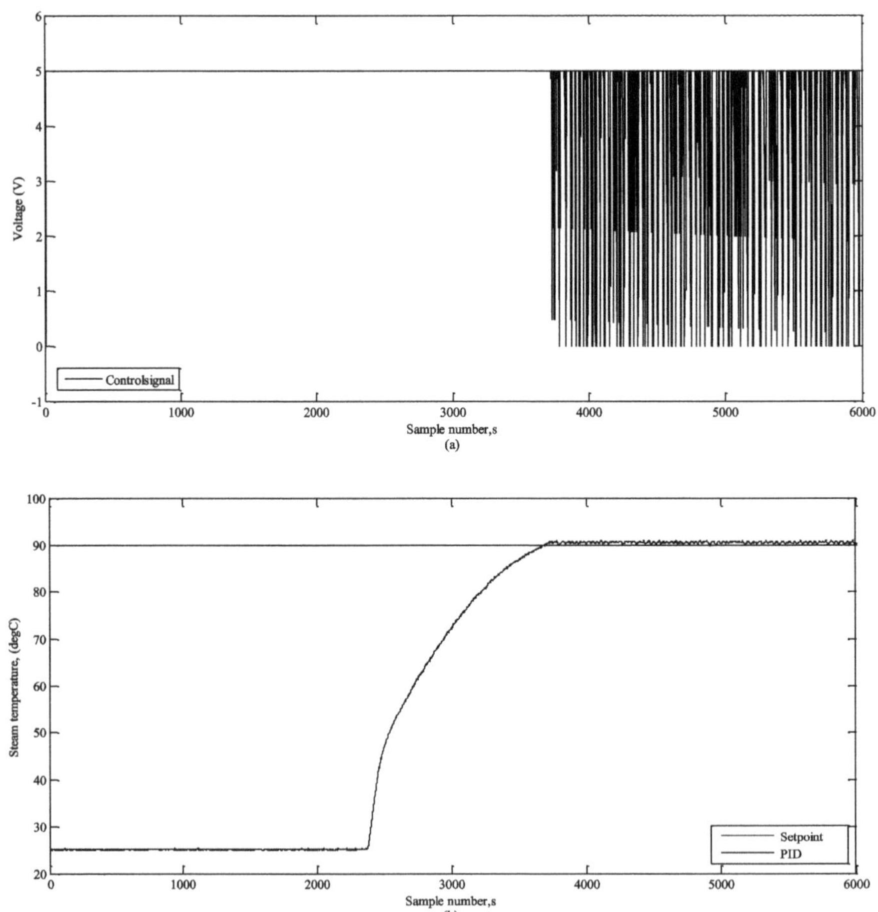

Fig. 7.1 Real-time performance for PID controller, (a) control signal, (b) output response

Table 7.1 Analysis performance for real-time PID controller using ZN method

No.	Controller	Rise time (s)	Settling time (s)	% OS	RMSE
1	PID	959	3657	0.3340	0.6034

7.3 Real Time of Step Response Performance for HFPID Controller

The performances of the HFPID controller scheme using different membership function are illustrated in Figs. 7.2, 7.3, and 7.4 [1]. Figure (a) presents the control signal, and Figure (b) presents the output response. The analysis results of performance evaluations are summarized in Tables 7.2, 7.3, and 7.4 for HFPID-3, HFPID-5 and HFPID-7, consecutively [1]. From the real-time implementation, it is quite revealing that the proposed HFPID controller scheme provides satisfactory temperature control due to its rise time, settling time and RMSE as compared with the conventional PID controller. Generally, the HFPID-3, HFPID-5, and HFPID-7 controllers settled at the desired temperature by provide faster response compared to PID controller. As expected, the HFPID-7 provides the most significant performance among all HFPID schemes in controlling steam temperature. Referring to Fig. 7.4, the HFPID-7 able to control the steam temperature in a fastest rise time, lowest settling time and also minimize the effect of fluctuation at a steady state (RMSE). From the control signal responses, it was observed that once the output response reached the desired set point, the control signal start varies between 0 and 5 V. The control signal sent to the heater was computed by compared the error at that time between the output response and desired set point.

7.4 Real Time of Step Response Performance for STFPID Controller

Figures 7.5, 7.6, and 7.7 show the step response performances for real-time implementation using STFPID-3, STFPID-5 and STFPID-7 controller in controlling and regulating the steam temperature in a hydro-diffusion system, respectively [1]. On the other hand, Tables 7.5, 7.6, and 7.7 tabulated the analysis performances for those respective controllers [1]. In general, the STFPID control techniques are shown to be adequate for essential oil extraction process and able to regulate the steam temperature at desired. The results show that STFPID controller offers better performance compared to PID controller on reducing rise time, settling time and minimized ripple at a steady state. From the step response results, it can be observed that the STFPID-5 produced the best performance and shows significant differences between STFPID-3 and STFPID-7. The RMSE however, show preferences on the STFPID-7, followed by the STFPID-5 and STFPID-3 with 0.5786, 0.3316, and 0.2286, respectively. Increase the number of membership functions will increase the steady-state error (RMSE).

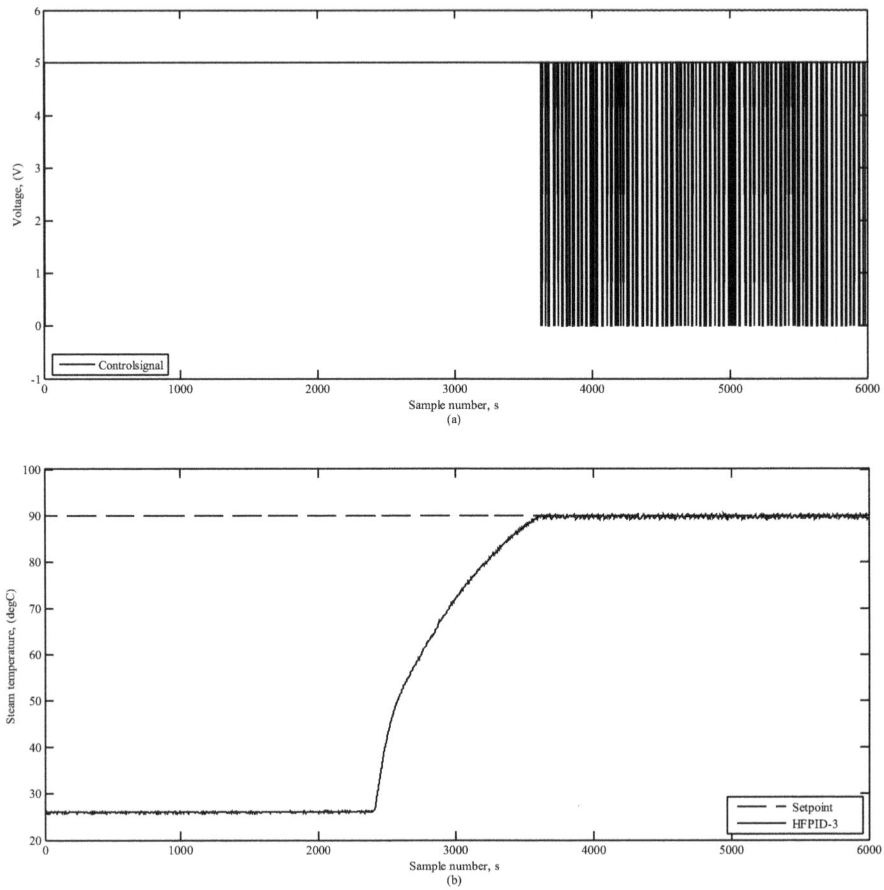

Fig. 7.2 Real-time performance for HFPID-3 controller

On the other hand, the control signal responses are consistent with those of PID and HFPID controller where it starts varies between 0 and 5 V when the output response achieves the desired set point.

7.5 Summary of Real-Time Step Response Performances

Figure 7.8 shows the comparison of step response for all developed controllers [1]. Meanwhile, Table 7.8 summarized the comparative analysis performance for real-time implementation of PID, HFPID-3, HFPID-5, HFPID-7, STFPID-3, STFPID-5, and STFPID-7 controller [1]. As can been seen from all the step response results, real-time experiments on controlling the steam temperature in an essential oil process are

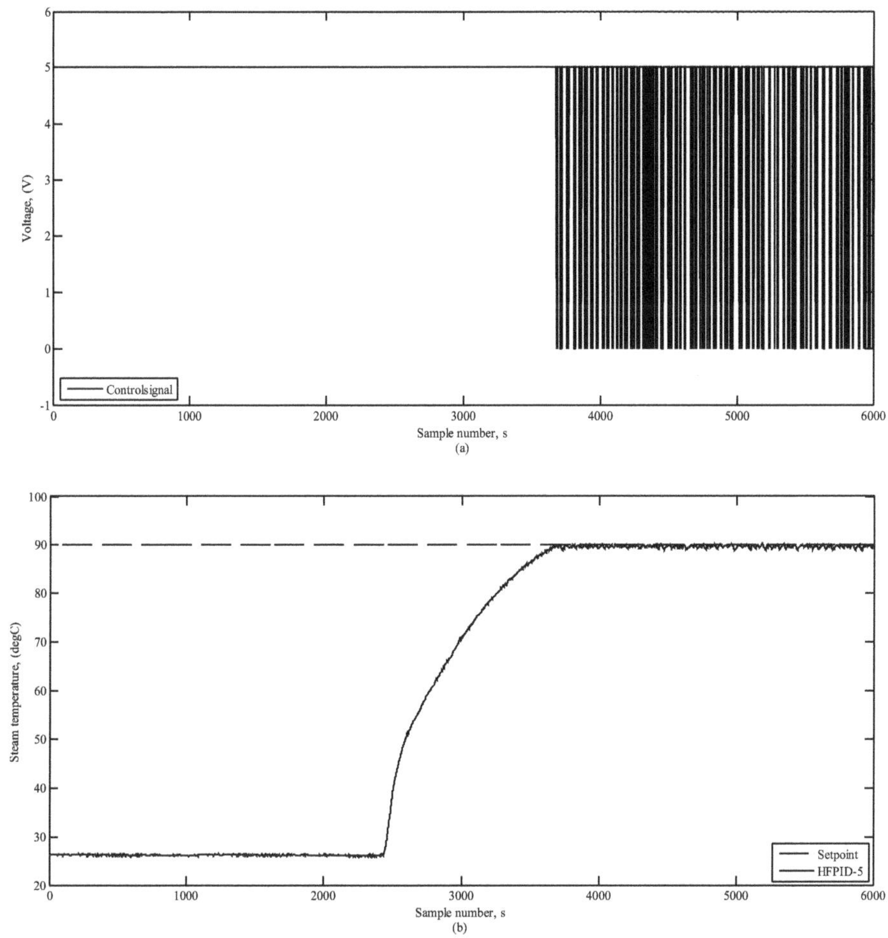

Fig. 7.3 Real-time performance for HFPID-5 controller

tested under long constant time delay, which is approximately 2400 s. The conventional PID, HFPID, and STFPID controller schemes are able to cope with the long-time delay through hydro-diffusion steam distillation system. In general, the intention to design controllers was achieved when all proposed controllers able to drive the steam temperature to the set point.

Figures 7.9, 7.10, 7.11, and 7.12 show the graphical comparisons of rise time, settling time, %OS and RMSE for real-time performances, respectively [1]. Referring to Figs. 7.9 and 7.10, it is apparent that the output response controlled by HFPID and STFPID controller is much faster than that controlled by PID controller, i.e. faster rise time and settling time. The HFPID-7 took the shortest rise time and settling time among all controllers with 836 s and 3500 s, respectively. It is clearly shown that the PID controller takes the longest time to achieve the desired set point which is

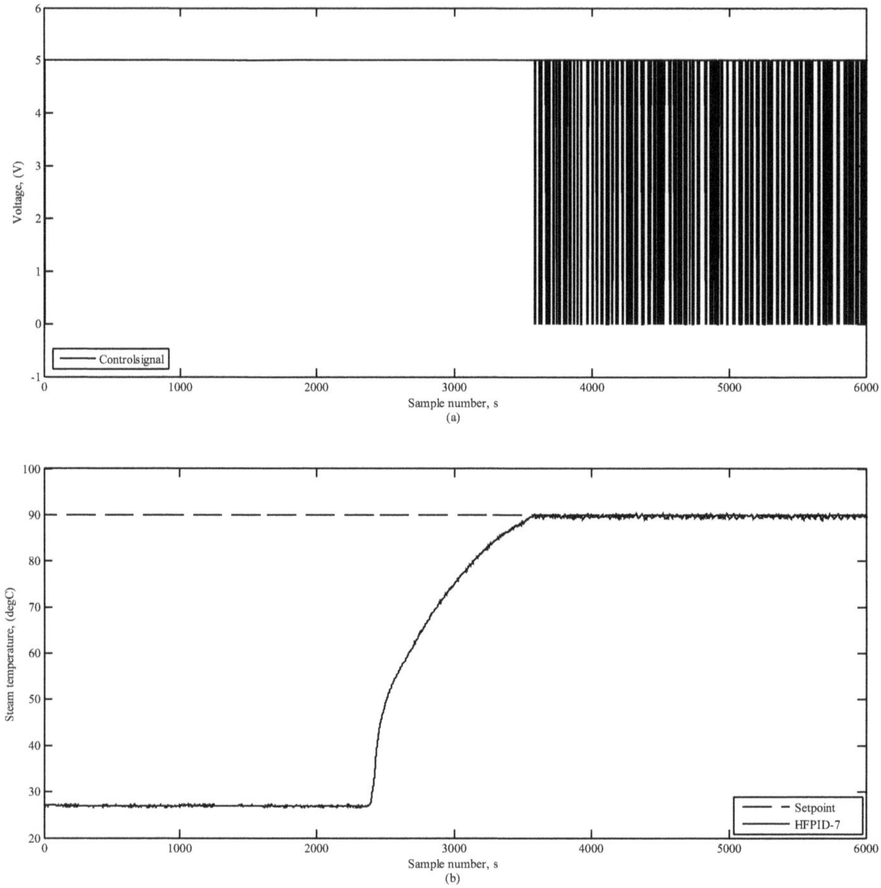

Fig. 7.4 Real-time performance for HFPID-7 controller

Table 7.2 Analysis performance for real-time HFPID-3 controller

No.	Controller	Rise time (s)	Settling time (s)	% OS	RMSE
1	HFPID-3	887	3581	0.5742	0.4404

Table 7.3 Analysis performance for real-time HFPID-5 controller

No.	Controller	Rise time (s)	Settling time (s)	% OS	RMSE
1	HFPID-5	879	3541	0.6749	0.3768

Table 7.4 Analysis performance for real-time HFPID-7 controller

No.	Controller	Rise time (s)	Settling time (s)	% OS	RMSE
1	HFPID-7	836	3500	0.7867	0.3656

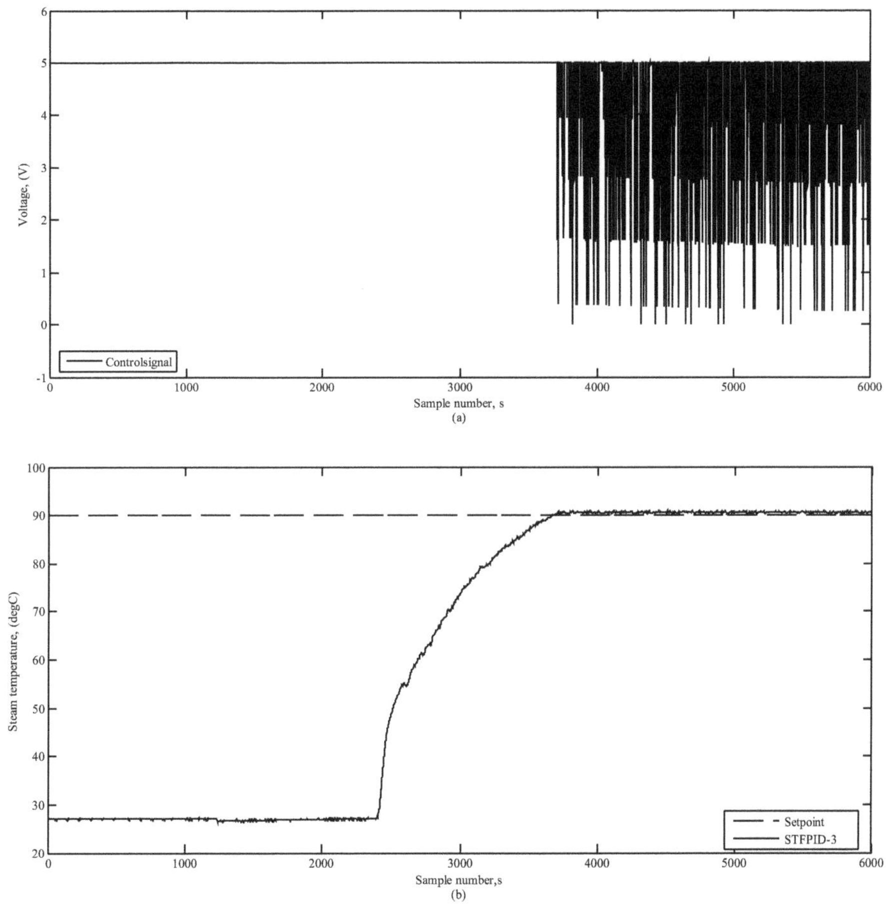

Fig. 7.5 Real-time performance for STFPID-3 controller

959 s. The PID controller also takes the longest time to settle down and maintain at a steady state which is 3657 s. This is more evident where the rise time for HFPID-7 is about 123 s (13%) less than that PID controller. The discrepancy is quite significant. Thus, the controller based on fuzzy logic is more significant to a system with large time delay. It pushes the output responses to attain a steady state much faster. Consequently, it will significantly reduce the power consumption. However, the conventional PID controller still performed relevant and acceptable rise time and settling time for the system that is categorized as long time delay.

Further analysis shows that for HFPID controller results, the increment in the number of membership function will significantly reduce the rise time and settling time. The HFPID-7 has generally produced the best rise time and settling time, followed by HFPID-5 and HFPID3 with (836, 3500 s), (879, 3541 s), and (887, 3581 s), respectively. As for STFPID controllers, it's contrary to expectations, the

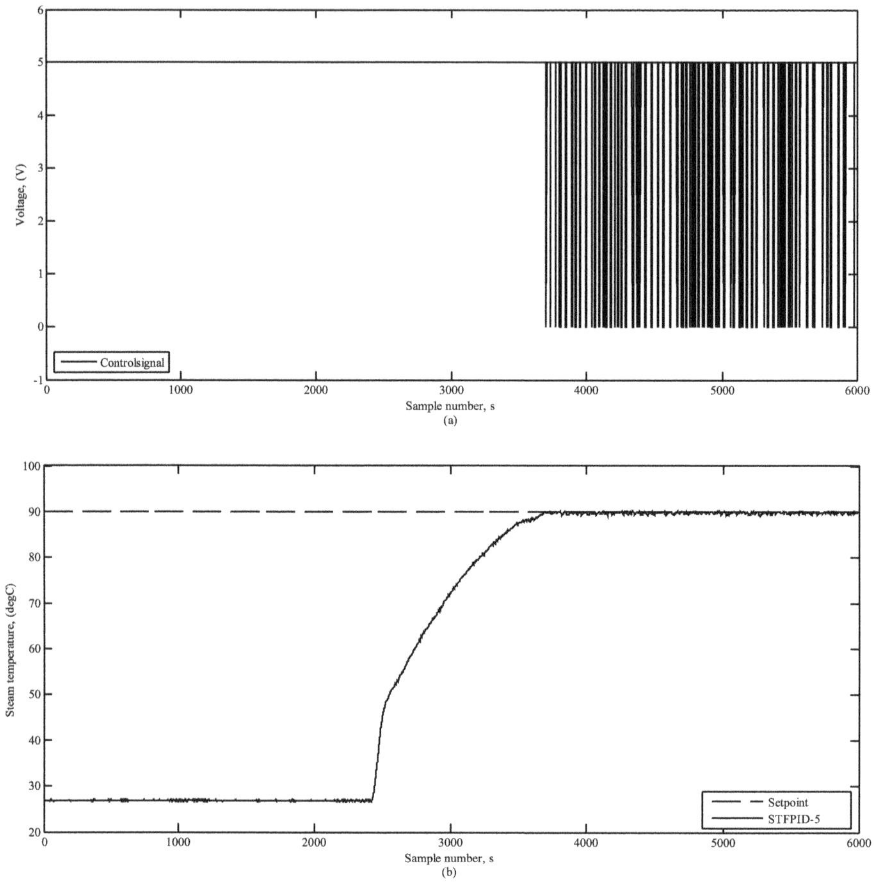

Fig. 7.6 Real-time performance for STFPID-5 controller

rise time is showing preferences on the STFPID-5 (882 s), followed by the STFPID-7 (906 s) and STFPID-3 (936 s). However, the discrepancies of settling time between STFPID controllers are not obvious. It means that for STFPID controller, increasing the membership functions will not have a great impact in terms of how fast the controller can perform.

Referring to Fig. 7.11, it is observed that the percentage of overshoot is shown to be the smallest for PID controller, which is 0.334%. Meanwhile, the %OS for HFPID-3, HFPID-5, HFPID-7, STFPID-3, STFPID-5, and STFPID-7 are 0.5742%, 0.6749%, 0.7867%, 0.4459%, 0.3364%, and 0.3364%, respectively. From this data, it is clearly the HFPID-7 shown to be the largest %OS. However, it's still acceptable because it shows less than 1% overshoot. It also can be observed that the STFPID-5 and STFPID-7 show comparable %OS. The %OS for HFPID-7 is a little bit higher than PID controller. The PID controller shows significant reduction of overshoot,

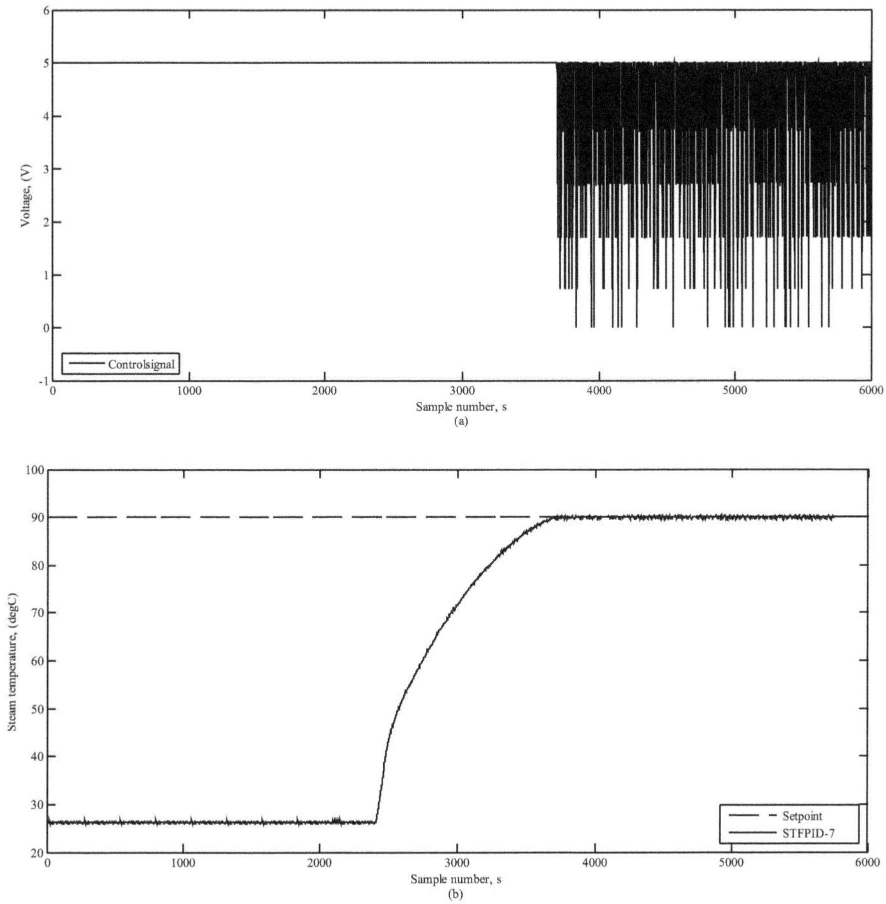

Fig. 7.7 Real-time performance for STFPID-7 controller

Table 7.5 Analysis performance for real-time STFPID-3 controller

No.	Controller	Rise time (s)	Settling time (s)	% OS	RMSE
1	STFPID-3	936	3639	0.4459	0.5786

Table 7.6 Analysis performance for real-time STFPID-5 controller

No.	Controller	Rise time (s)	Settling time (s)	% OS	RMSE
1	STFPID-5	882	3634	0.3364	0.3316

Table 7.7 Analysis performance for real-time STFPID-7 controller

No.	Controller	Rise time (s)	Settling time (s)	% OS	RMSE
1	STFPID-7	906	3637	0.3364	0.2286

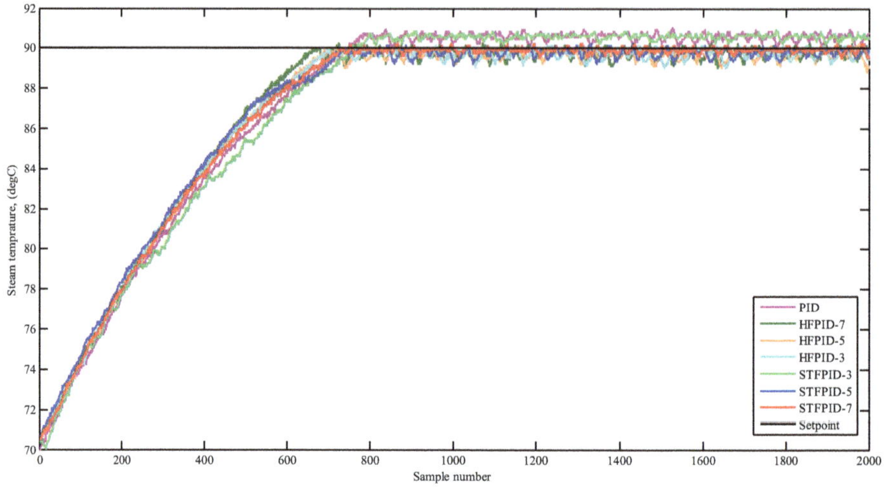

Fig. 7.8 Real-time performances of PID, HFPID-3, HFPID-5, HFPID-7, STFPID-3, STFPID-5, and STFPID-7 controller for steam temperature control

Table 7.8 Analysis performance for real-time implementation of PID, HFPID-3, HFPID-5, HFPID-7, STFPID-3, STFPID-5, and STFPID-7 controller

No.	Controller	Rise time (s)	Settling time (s)	% OS	RMSE
1	PID	959	3657	0.3340	0.6034
2	HFPID-3	887	3581	0.5742	0.4404
3	HFPID-5	879	3541	0.6749	0.3768
4	HFPID-7	836	3500	0.7867	0.3656
5	STFPID-3	936	3639	0.4459	0.5786
6	*STFPID-5*	*882*	*3634*	*0.3364*	*0.3316*
7	STFPID-7	906	3637	0.3364	0.2286

The italic values are the best result between all controllers

which is 0.4527% less than HFPID-7. Referring to Fig. 7.9, it's implied that faster rise time will produce large overshoot in the output response.

Analyzing the RMSE results for real-time implementation, it shows that all controllers produce a satisfactory result since the RMSE values for all controllers are below 1, which is considered as a small value. However, the RMSE results present in Fig. 7.12 generally agree on the rise time and settling time results where the RMSE value for PID controller is slightly higher than other controllers. The PID controller's RMSE value is 0.6034. The PID controller produced highest oscillation before reach the stable region compared with other controllers.

Referring the RMSE values for HFPID controller schemes, show preferences on the HFPID-7 with 0.4404, followed by HFPID-5 with 0.3768 and HFPID-3 with 0.3656. Overall, based on the fluctuations observed from the output response, the

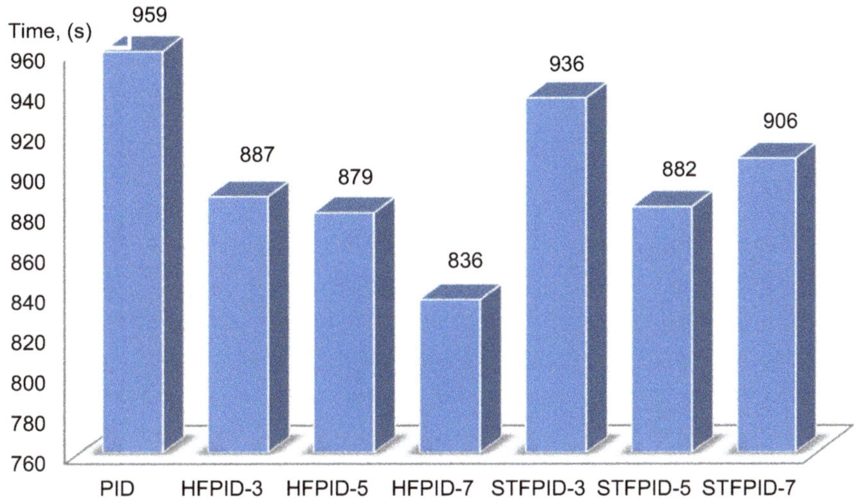

Fig. 7.9 Rise time of real-time implementation for all controllers

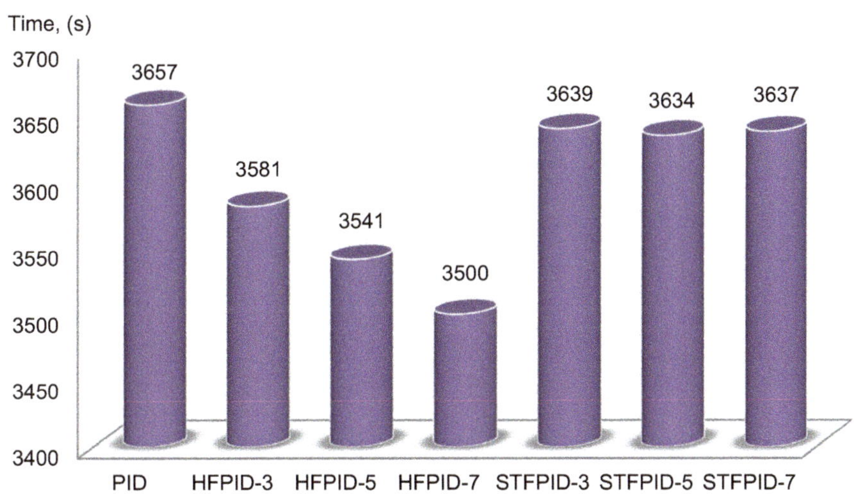

Fig. 7.10 Settling time of real-time implementation for all controllers

RMSE results show that the STFPID-7 has advantages over all controllers since it produces the smallest value of RMSE, which is 0.2286. The output response of STFPID-7 is very precise and seem adequately fulfilled the control objective with small value of deviation. The STFPID-7 shows significant discrepancies of RMSE, which is about 0.3748 less than PID controller whereas, the RMSE values reduced significantly from 0.5786 to 0.3316 for STFPID-3 and STFPID-5, respectively. As

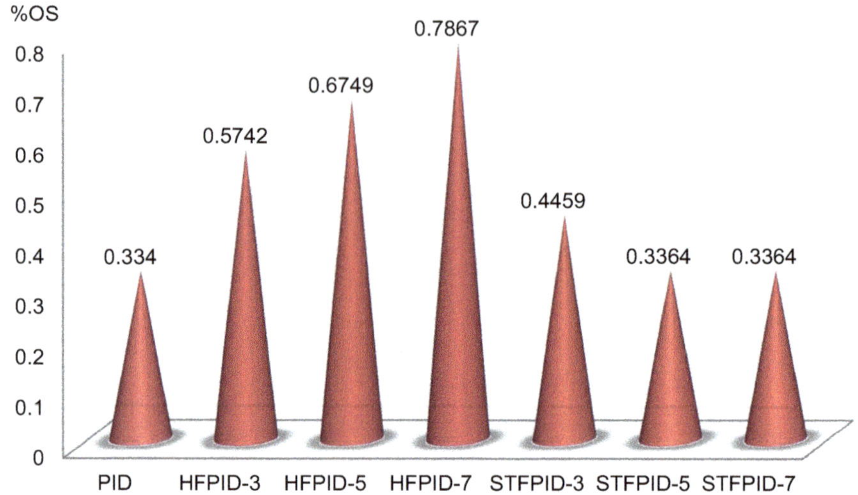

Fig. 7.11 Percentage overshoot of real-time implementation for all controllers

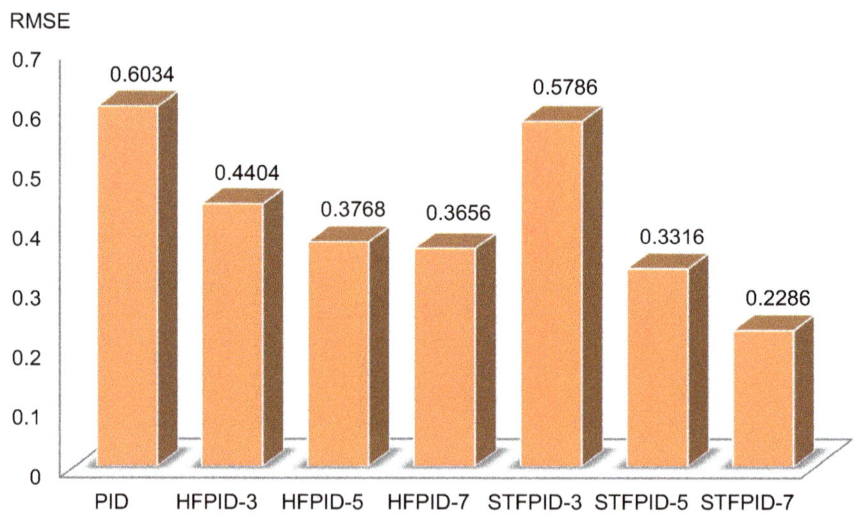

Fig. 7.12 RMSE of real-time implementation for all controllers

shown in Fig. 7.12, this finding as expected where increase the membership function will improve the RMSE value.

The purposed of this study to implement the HFPID and STFPID controllers to control the steam temperature in hydro-diffusion steam distillation essential oil extraction system was achieved when HFPID and STFPID show better performances as compared with PID controller. Overall, from the step response performances, the controllers based on fuzzy logic (HFPID and STFPID) produces small rise time, fast settling time, acceptable %OS, and small steady-state error. As long as the rise time, settling time, %OS, and fluctuations of the output response are small, the stability can be maintained. Next, the effectiveness of developed controllers was tested by performed robustness analysis. For robustness test, PID, HFPID-7, and STFPID-5, controllers were implemented and the performances were compared. The HFPID-7 and STFPID-5 controllers were chosen since produces the best performances among HFPID and STFPID controllers.

7.6 Robustness Test for Real-Time Implementation

Two tests were carried out, which are set point tracking and introduce load disturbance. The aim of this study is to analysis the robustness of each controller in adapting the disturbance. Robust controller literally means the controller should remain stable in all situations even when set point changes or suddenly introduce disturbance during running process.

7.6.1 Set Point Tracking

This study is to analyze the ability of conventional PID controller and to evaluate the degree of improvement given by HFPID and STFPID controller as compared to conventional PID controller in tracking the set point during running process. Three desired set point were set, which are 60 °C, 80 °C, and 90 °C, respectively. Each set point setting provides the different characteristics. Each proposed controller were analyzed based on rise time, settling time, %OS, and RMSE for each desired set point (Tables 7.9 and 7.10; Figs. 7.13 and 7.14).

Figure 7.15a is the control signal, and Fig. 7.15b is the output response illustrates the real-time performance of STFPID-5 in controlling the system using step change input [1]. Meanwhile, Table 7.11 summarized the set point change performance for STFPID-5 controller [1]. From the result, it shows that the STFPID-5 able to regulate the steam temperature in tracking the changes of set point. The rise time performances are 129, 245 and 206 s at a set point 60, 80, and 90 °C, consecutively. The %OS at a set point 90 °C shows an excellent result that recorded only 0.3249% as compared to

Table 7.9 Analysis performance for real-time PID controller on set point tracking [1]

No.	Temperature (°C)	Rise time (s)	Settling time (s)	% OS	RMSE
1	80	191	1594	1.1108	0.7129
2	85	328	1495	0.7462	0.4087
3	90	334	1496	0.6723	0.3135

Table 7.10 Analysis performance for real-time HFPID-7 controller on set point tracking [1]

No.	Temperature (°C)	Rise time (s)	Settling time (s)	% OS	RMSE
1	60	131	1591	0.9110	0.7703
2	80	247	1495	0.5047	0.5566
3	90	216	1497	0.3371	0.2851

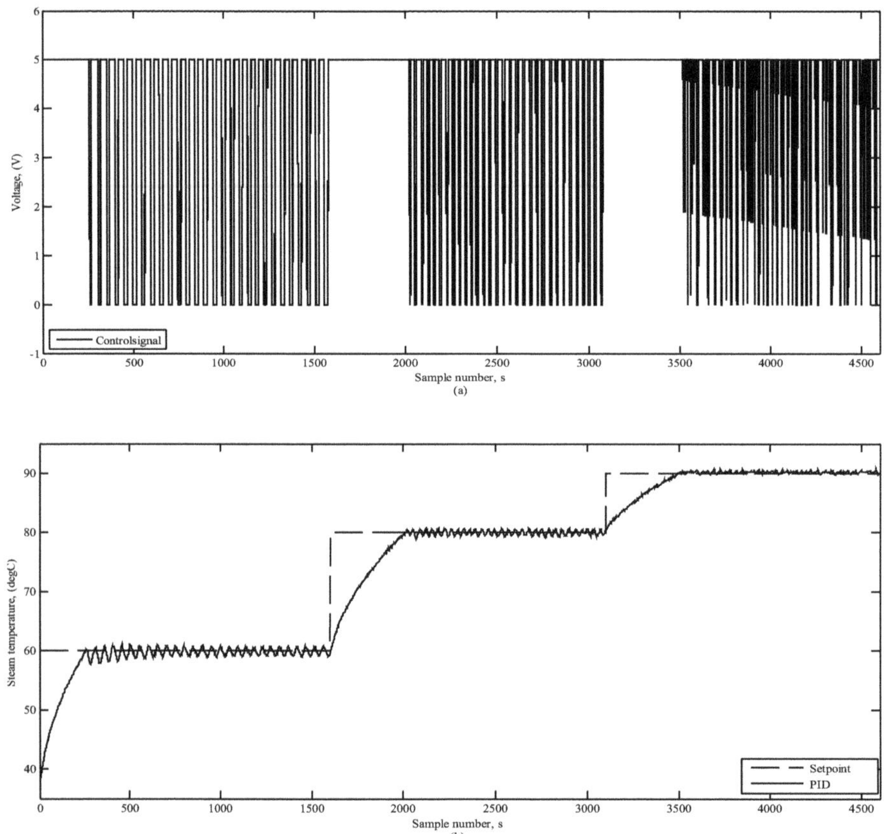

Fig. 7.13 Real-time PID controller performance on set point tracking [1]

the set point 60 °C with 0.8354% and 80 °C with 0.4953%. By analyzing the RMSE result, it found that the output response at the set point 90 °C is very precise with smallest RMSE value, which is 0.3040. The RMSE value for set point 80 and 90 °C are 0.7372 and 0.5677, respectively.

Figure 7.16 depicts the combination result for set point change performances using PID, HFPID-7 and STFPID-5 controller [1]. From the set point change test, all developed controllers are able to track the changes of set point, whether in small or large set point change. The PID, HFPID-7, and STFPID-5 controller produces

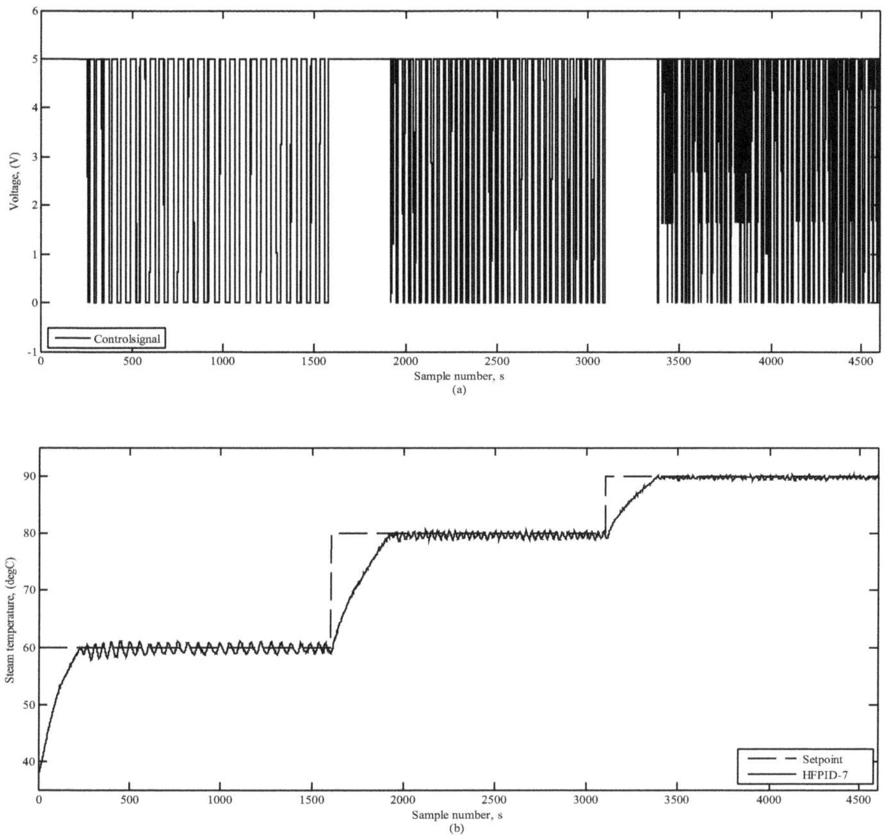

Fig. 7.14 Real-time HFPID-7 controller performance on set point tracking [1]

a satisfactory result. However, the results show that HFPID-7 and STFPID-5 offer better performances, especially on providing fast rise time, settling time and less overshoot to compare to PID controller. The HFPID-7 and STFPID-5 precisely enhances its performance when facing set point changes. This improvement generally will lead to faster extraction process. However, the PID controller provides comparable performance of RMSE value as HFPID-7 and STFPID-5.

7.6.2 Load Disturbance

Figures 7.17, 7.18, and 7.19 depicted the results for real-time performance of PID, HFPID-7, and STFPID-5 controller on recovering load disturbance, accordingly [1]. Meanwhile, Fig. 7.20 shows a comparison performance among different control strategies for steam temperature on hydro-diffusion steam distillation system with

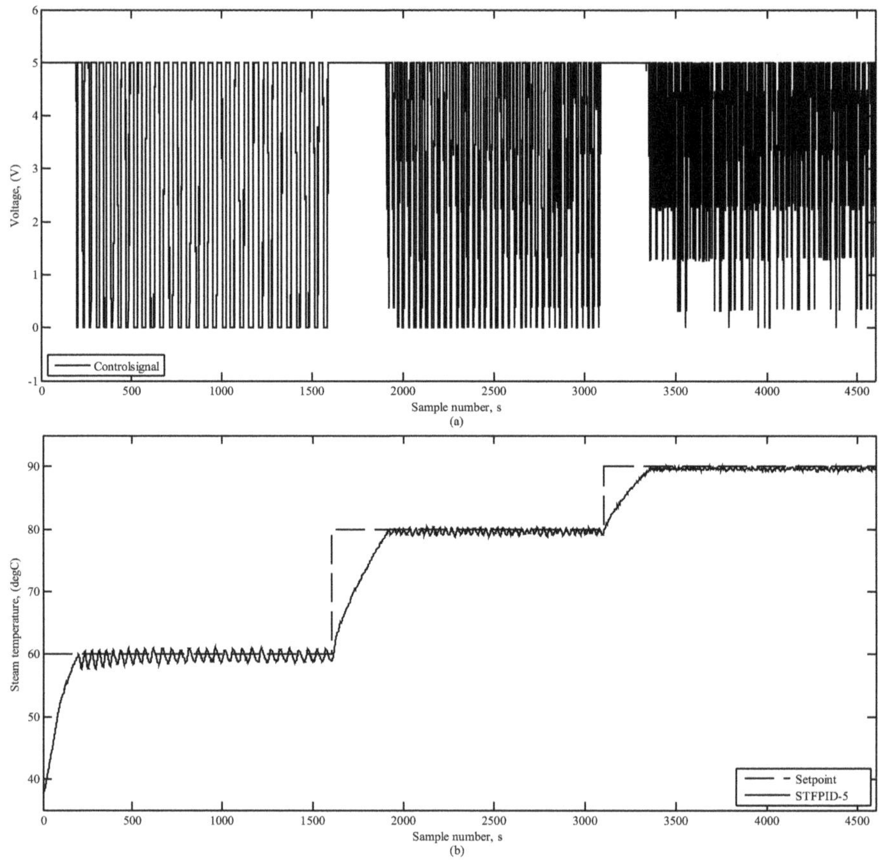

Fig. 7.15 Real-time STFPID-5 controller performance on set point tracking

Table 7.11 Analysis performance for real-time STFPID-5 controller on set point tracking

No.	Temperature (°C)	Rise time (s)	Settling time (s)	% OS	RMSE
1	60	129	1581	0.8354	0.7372
2	80	245	1484	0.4953	0.5677
3	90	206	1478	0.3249	0.3040

load disturbance [1]. Robustness of the controller design will be based on the how a fast output response tackles back the desired set point.

The result for real-time of PID controller on recovering load disturbance for steam temperature control is shown in Fig. 7.17a control signal and Fig. 7.17b output response [1]. The result indicates that the PID controller able to returns to the set point by taken 386 s after injecting the disturbance at 1000th samples.

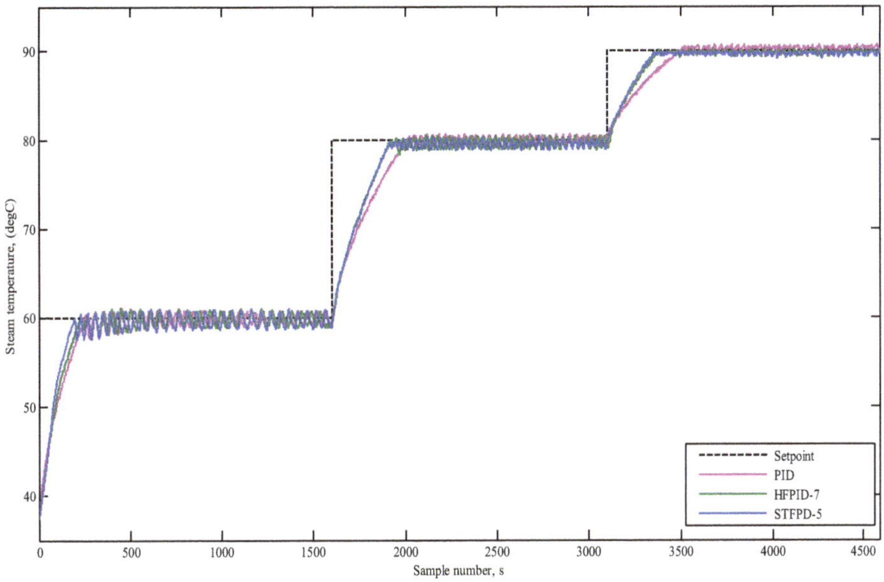

Fig. 7.16 PID, HFPID-7, and STFPID-5 performances on setpoint tracking

Figure 7.18a control signal and Fig. 7.18b output response depicted the result for real-time implementation of hybrid fuzzy plus PID controller using 7 membership functions on recovering load disturbance test [1]. From the result, it can be noted that the HFPID-7 scheme shows a satisfactory result and able to returns to the set point by taken 74 s after injecting the disturbance.

The result of self-tuning fuzzy plus PID controller using 5 membership functions on recovering load disturbance is shown in Fig. 7.19 [1]. The result demonstrates that the proposed STFPID-5 scheme take 101 s to return to the desired set point (90 °C) when load disturbance is suddenly added at 1000th samples.

Table 7.12 summarized the results for real-time performance of PID, HFPID-7, and STFPID-5 controller on recovering load disturbance whereas Fig. 7.21 shows the graphical comparison of recovery time for all the proposed controllers [1]. From the result in the Table 7.12, after introducing the disturbance at sample 1000th, steam temperature decrease with $T_{\min} = 87.42$ °C, $T_{\min} = 87.50$ °C, and $T_{\min} = 87.42$ °C for proposed PID, HFPID-7 and STFPID-5 controllers, respectively. All controllers are able to capture back the set point after the system was disturbed during running process. However, these results revealed that the good robustness of the HFPID-7 and STFPID-5 controller since PID controller took the longest time on recover load disturbance with 386 s. The time taken to capture back the set point after introduce disturbance for HFPID controller using 49 rules is five times faster, which is only 74 s as compared with PID controller with 386 s. The recovery process is significantly shortened for HFPID-7 by taken 312 s less than PID controller.

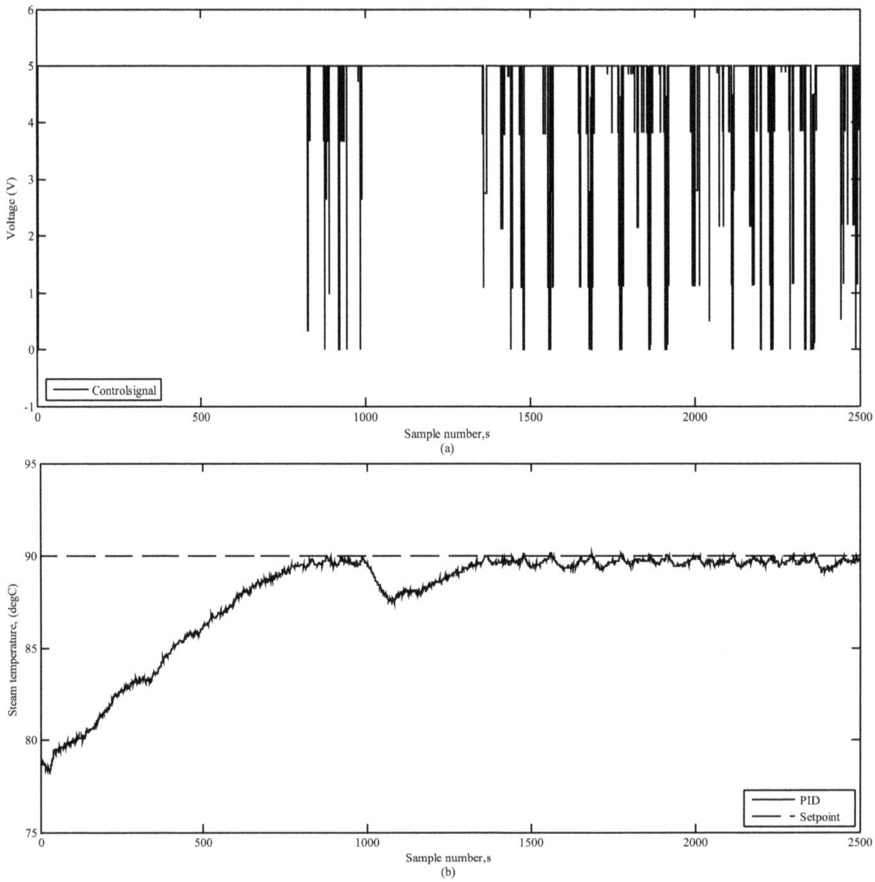

Fig. 7.17 Real-time PID controller performance on recovering load disturbance

Significant improvement can be spotted to HFPID-7 controller because provide speedy recovery and robust response in comprising an abrupt disturbance during running process. The results also demonstrate that the STFPID-5 offer better performance by greatly improve the recovery process compared with PID controller. The STFPID-5 takes 285 s less than PID controller. On the other hand, the control signals for all proposed controllers to display similar pattern and very uniform. The control signal varies between 0 and 5 V at a steady state. However, when subjected to load disturbance at 1000th samples, the control signal maintained at 5 V. The developed controllers pushed the steam temperature to return to the set point as fast as possible.

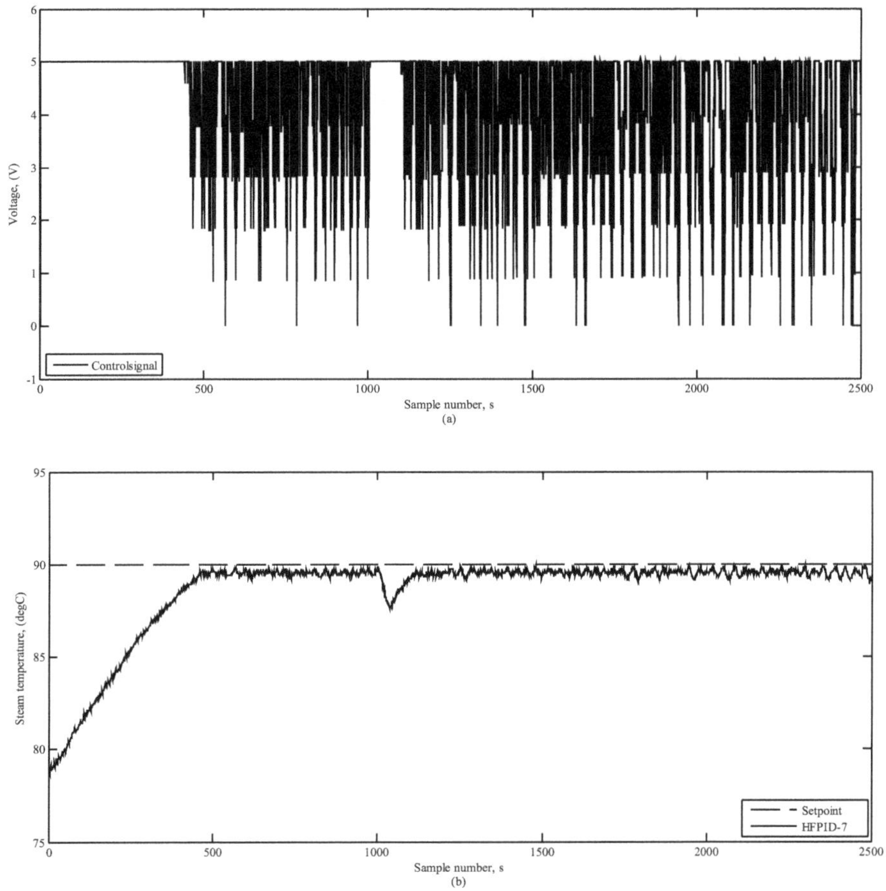

Fig. 7.18 Real-time HFPID-7 controller performance on recovering load disturbance

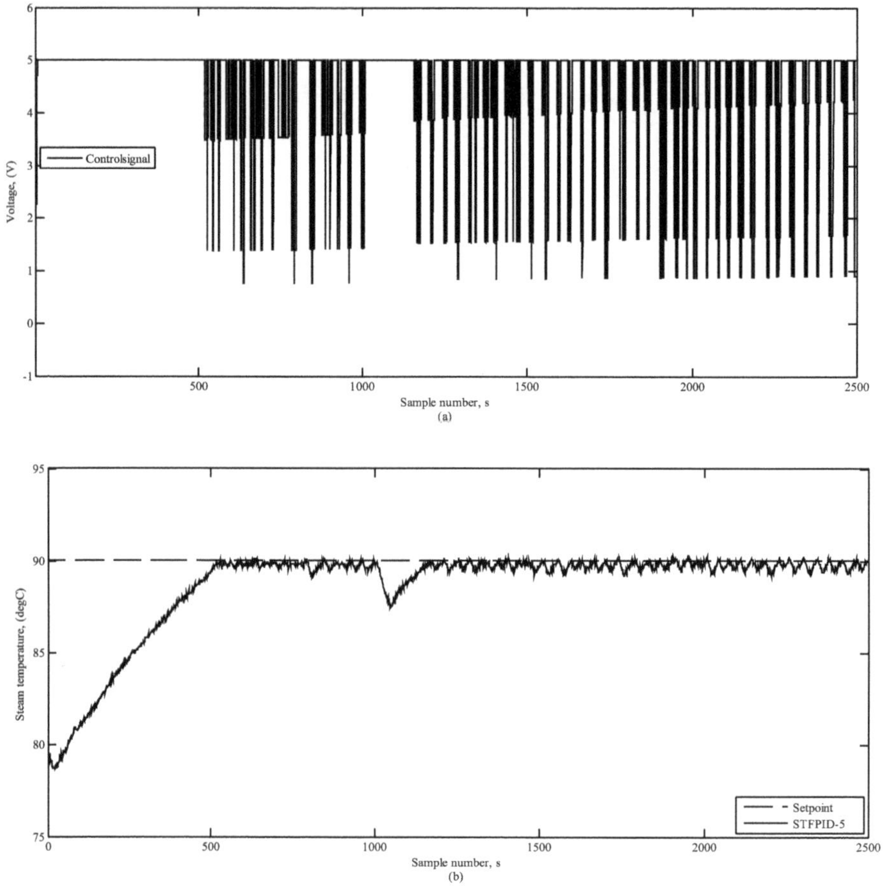

Fig. 7.19 Real-time STFPID-5 controller performance on recovering load disturbance

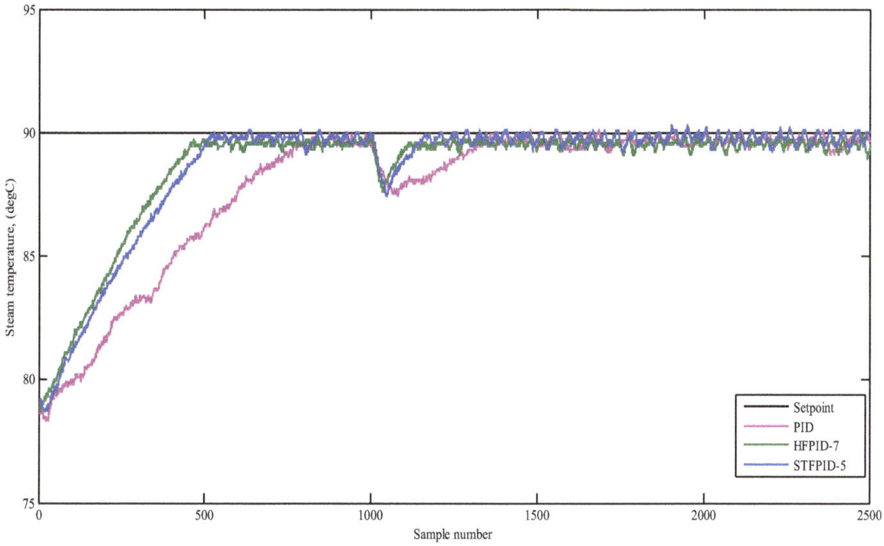

Fig. 7.20 PID, HFPID-7 and STFPID-5 performance on recovering load disturbance

Table 7.12 Analysis for real-time performance of PID controller, HFPID-7, and STFPID-5 controller on recovering load disturbance

No.	Controller	T_{min}	Recovery time (s)
1	PID	87.42 °C	386
2	HFPID-7	87.50 °C	74
3	STFPID-5	87.42 °C	101
HFPID-7 compared with the PID			– 312
HFPID-7 compared with the STFPID-5			– 27
STFPID-5 compared with the PID			– 285

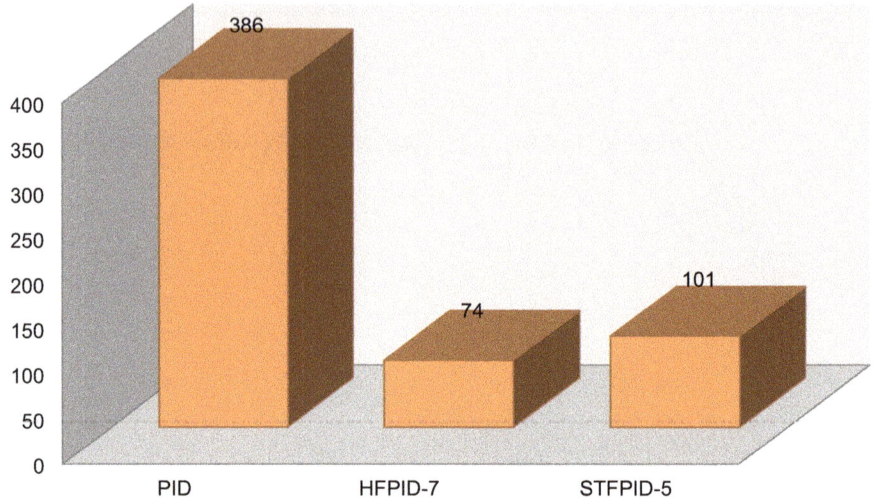

Fig. 7.21 Graphical comparison of recovery time for all proposed controllers

Reference

1. Yusoff, Z.M.: Self-tuning Fuzzy PID Control of Hydro-diffusion Essential Oil Extraction System. Universiti Teknologi MARA, Malaysia (2014)

Chapter 8
Extraction of Kaffir Lime Essential Oil Using Intelligent Controller

Abstract This chapter presents the practical application of intelligent control strategies for the extraction of kaffir lime essential oil using a hydro-diffusion steam distillation system. Building upon the real-time validation of advanced control architectures in Chap. 7, the validated Hybrid Fuzzy-PID (HFPID), and Self-tuning Fuzzy-PID (STFPID) controllers are implemented to enhance thermal regulation during the extraction process. Precise steam temperature control plays a critical role in preserving volatile compounds, improving oil yield, and maintaining product quality. Experimental results demonstrate that intelligent controllers significantly enhance temperature stability, reduce fluctuation, and improve system responsiveness compared to conventional PID control. The improved thermal consistency contributes directly to more efficient extraction performance and better energy utilization. The findings confirm that intelligent fuzzy-based control is not only theoretically superior but also practically beneficial in agro-industrial applications. This study establishes a validated framework for integrating advanced control techniques into essential oil extraction systems, supporting the development of smarter, more efficient, and quality-driven botanical processing technologies.

8.1 Sample Preparations

Fruits of kaffir limes were purchased from orchards around Dengkil area. The fresh peels of kaffir lime were put at room temperature around 23 °C, and it was protected from direct light. Then, peels of kaffir lime undergo weighing process. A 0.35 kg grounded kaffir lime peels for each steam temperature setting, which are 80, 90, 95 °C, and uncontrolled temperature, which is 100 °C. The steam temperature setting was achieved using a HFPID controller scheme by controlling the voltage fed to the heater. Peels have much aromatic sense, and much moisture compared to leaves, many contributions to oil production in steam distillation. The yield of the essential oil produced was estimated on a wet weight basis. Figure 8.1 depicted the work done for sample preparations in extracting the kaffir lime [1].

Z. Mohd Yusoff et al., *Fuzzy Logic, Smart Control, and the Future of Essential Oils*, Studies in Fuzziness and Soft Computing 442,
https://doi.org/10.1007/978-981-92-2084-7_8

Fig. 8.1 Sample preparations

8.2 Extraction Methods

The extraction was done by using a developed hydro-diffusion steam distillation system. By using the hydro-diffusion system, the peels were placed in the material tray inside hydro-diffusion steam distillation plant and go for extraction process. Figure 8.2 shows the extraction process using hydro-diffusion steam distillation system [1]. Extraction was performed approximately 2 h because based on the previous work for 0.35 kg, the extraction was performed for 2 h. Otherwise not much essential oil produces over 2 h and save energy consumption. The collected essential oils were separated from water by using sodium anhydrous acid and dichloromethane, which acted as catalysts using Vacuum Rotavap. The genuine oil is then stored in sealed containers under refrigeration prior to analysis.

To measure the quantity of the collected oils, the percentage yields of essential oil were calculated as follows [1]:

Fig. 8.2 Extraction process using hydro-diffusion steam distillation system

$$\text{Yield}(\%) = \frac{\text{Weight of extracted recovered}}{\text{Weight of fresh citrus peel}} \times 100 \tag{8.1}$$

where the oil density of kaffir lime oil is 0.871 g/mL. So, the weight of extract recovered in grams can be derived as [1]:

$$\begin{aligned}\text{Weight of extract recovered(in gram)} &= \text{volume of oil collected(mL)} \\ &\quad \times \text{density of kaffir lime oil}\left(\frac{0.871\ \text{g}}{\text{mL}}\right)\end{aligned} \tag{8.2}$$

8.3 Analysis of Essential Oils

The composition of kaffir lime oil was tested by GC and GC/MS as shown in Fig. 8.3 for qualitative analysis to identify active constituents [1]. For GC analysis, a Shimadzu GC 2010 Gas Chromatograph which equipped with an FID detector and fused silica capillary column CBP5 (30 m × 0.25 mm, 0.25 µm film thickness) was employed. The GC/MS analyses were carried out by using Agilent 5975C Mass Selective Detector equipped with an Agilent 7890A Gas Chromatograph. An HP-5MS fused silica capillary column (30 m × 0.25 mm) with 0.25 µm film thickness was used. The carrier gas for GC is He with flow rate 1 ml/min and temperature programming as tabulated in Table 8.1 [1]:

The identification of the chemical compounds was done by matching their mass spectra with existing data in the HPCH2205. L, Wiley7 Nist05.L and NIST05a. L library searched data. Their retention indices were compared with Adam, 2007 libraries. Then, the major compounds in the kaffir lime were compared with those reported in literature.

Fig. 8.3 GC-MS analysis

Table 8.1 Setting for GC-MS analysis

Injector temperature	250 °C
Detector temperature	280 °C, with mass range recorder from 45–450 mass-to ratios, with electron energy of 70 eV
Oven temperature	The column oven was set to 60 °C for 3 min, and then the temperature was gradually increased up to 230 °C at a rate 3 °C/min. For 230 °C till 325 °C, the rate was maintained at 30 °C/min

8.4 Comparative Analysis on Effect of Temperature on the Production Rate of Kaffir Lime Oil

Figure 8.4 presents the production yield of kaffir lime using hydro-diffusion steam distillation system. The extraction process was performed for 2 h [1]. From the experiment, the production yield start produced approximately after 50 min. As shown in Fig. 8.4, two layers form after the extraction process where the top layer is hydrosol and water at the bottom of the container. The oils were collected for every 10 min with the objective to see the flow rate of production yield. From the Fig. 8.4, it is interesting to note that the yield of essential oil is rapidly produced at the early stage of extraction as shown in container 1. The yield in container 1 was collected after 50 min process started. Another important finding was that the yields start decreasing as time increasing. As shown in container 5, a poor amount of yield was collected after 2 h.

Table 8.2 tabulated the accumulated amount of percentage yield for each steam temperature control based on the wet weight basis [1]. It means that no drying process involved during sample preparations stage.

Fig. 8.4 Production yield of kaffir lime peel

Table 8.2 Percentage of extracted kaffir lime essential oil

Steam temperature (°C)	Production yield (mL)	Percentage yield (w/w)
80	2.8	0.70
90	8.8	2.19
95	11.8	2.94
Uncontrolled	12.5	3.11

Refer to Table 8.2, by using hydro-diffusion steam distillation system for 2 h, the percentage of kaffir lime oil yields at temperatures of 80 °C, 90 °C, 95 °C, and uncontrolled temperature were 0.70% (w/w), 2.19% (w/w), 2.94% (w/w), and 3.11% (w/w), respectively. It was notable that the percentage yield increased as steam temperature increased. Uncontrolled steam temperature yielded the highest amount of essential oil. However, the observed difference between uncontrolled steam temperature and steam temperature at 95 °C was not too significant. The lowest yield was obtained at steam temperature 80 °C. The most interesting finding was that the oil yield quite lucrative at 90 °C and showed 77.2% significant difference as compared to oil yield at 80 °C. The graphical comparison of percentage yield at various steam temperatures as a function of time is shown in Fig. 8.5 [1].

Figure 8.6 presents the production curves of kaffir lime oil yield at four different steam temperatures where the oil yield was collected for every 10 min [1]. The curves show that the extraction yield seems to have an asymptotic shape. It is apparent in this figure that oil yield was aggressively produced at early stage of the extraction process

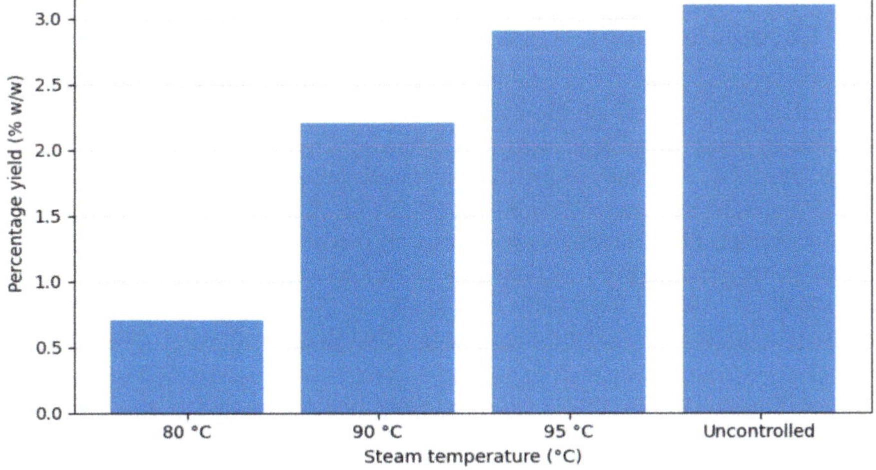

Fig. 8.5 Graphical comparison of percentage yield for extracted kaffir lime using hydro-steam distillation system

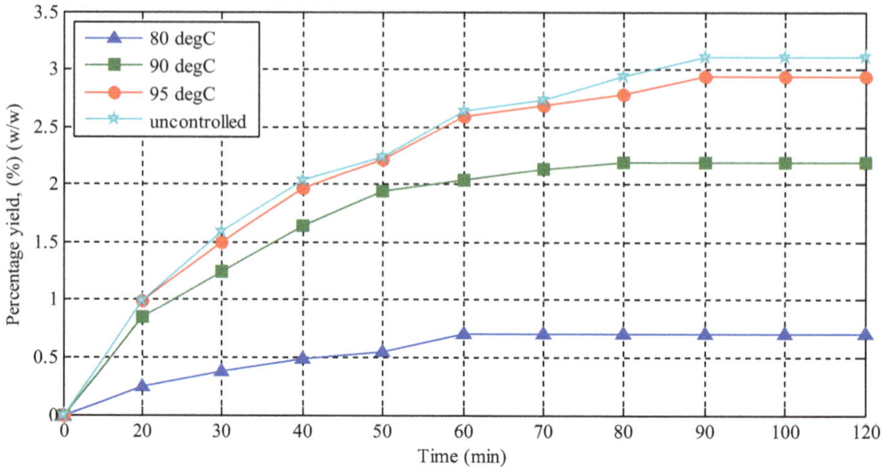

Fig. 8.6 Production curves of kaffir lime oil yield at various steam temperature control

for all steam temperature control. From the result, there are 3 different saturation times, which are 60 min, 80 min, and 95 min for temperature of 80 °C, 90 °C, 95 °C, and uncontrolled temperature, respectively. It was found that 80 °C has the fastest saturation time among all but produced the smallest oil yield. An extraction time of 60 min at 95 °C provided 3 times more yields as compared to those obtained by 80 °C.

8.5 Analysis of Kaffir Lime Essential Oil by Gas Chromatography–Mass Spectrometry (GC-MS)

The analysis of essential oil from kaffir lime peels for each steam temperature was done by using Gas Chromatography–Mass Spectrometry (GC-MS). The chromatograms of GC-MS analysis for all temperature studies are presented in Figs. 8.7, 8.8, 8.9 and 8.10 for temperature 80 °C, 90 °C, 95 °C, and uncontrolled temperature, respectively [1]. The relative peak areas in percentage (%) were calculated by comparing the peak ratio in chromatograms of GC and GC-MS. The compounds of essential oil obtained from kaffir lime peels for different temperature profile are summarized in Table 8.3. All the compounds were listed based on the retention index (RI) [1].

It can be seen from the data in Table 8.3 that a total of 26 compounds were found in kaffir lime peels extracted using hydro-diffusion steam distillation system start with α-Thujene (0924) until δ-Cadinene (1522). Different numbers of compounds were extracted where a total of 19 compounds, 24 compounds, 19 compounds

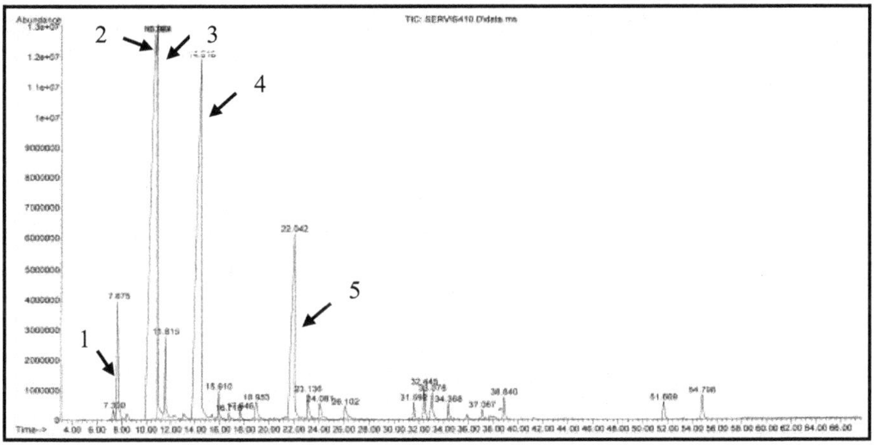

Fig. 8.7 GC-MS chromatogram of volatile compound from kaffir lime peel essential oil at 80 °C. The major compounds marked as (1) α-pinene, (2) sabinene, (3) β-pinene, (4) limonene, (5) citronellal

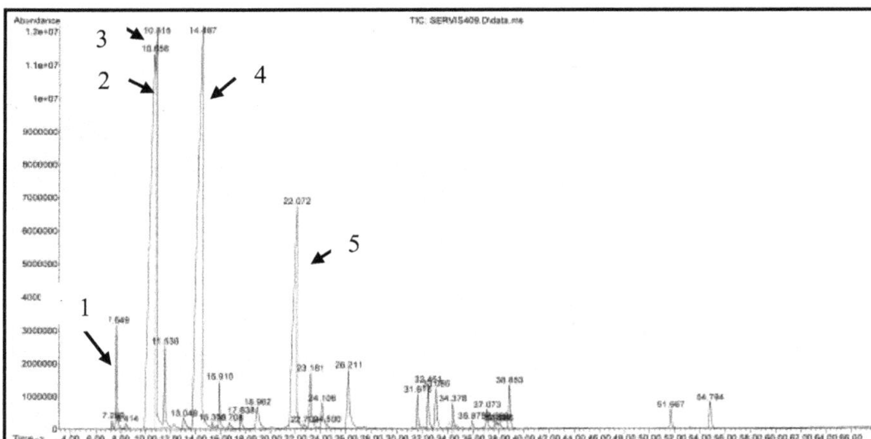

Fig. 8.8 GC-MS chromatogram of volatile compound from kaffir lime peel essential oil at 90 °C. The major compounds marked as (1) α-pinene, (2) sabinene, (3) β-pinene, (4) limonene, (5) citronellal

and 24 compounds were identified in temperature 80, 90, and 95 °C and uncontrolled temperature, accordingly. Therefore, the temperature at 90 °C and uncontrolled temperature can extract more chemical compounds. It is apparent from this table that 17 compounds appear in all temperature conditions such as α-Thujene, α-Pinene, sabinene, β-Pinene, mycrane, δ-2-Carene, limonene, ɣ-Terpinene, linalool, citronellal, terpinene-4-ol, α-Terpineol, citronellol, α-Copaene, E-caryophyllene, citronellyl propanoate, and germacrene D. All the major compounds of the kaffir

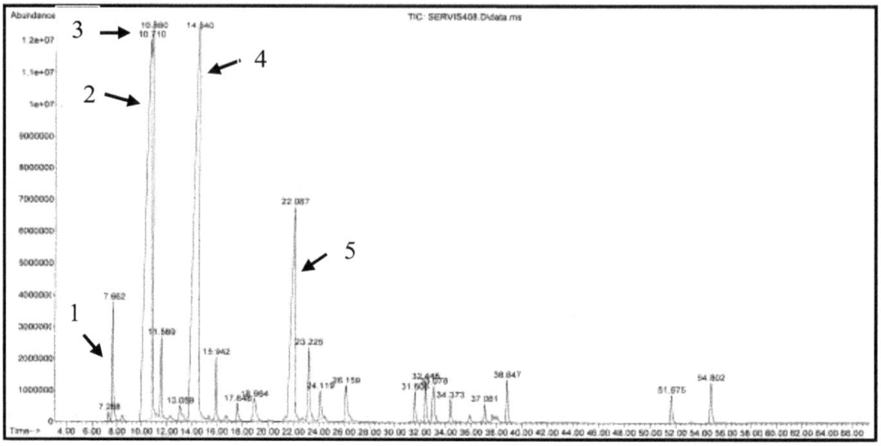

Fig. 8.9 GC-MS chromatogram of volatile compound from kaffir lime peel essential oil at 95 °C. The major compounds marked as (1) α-pinene, (2) sabinene, (3) β-pinene, (4) limonene, (5) citronellal

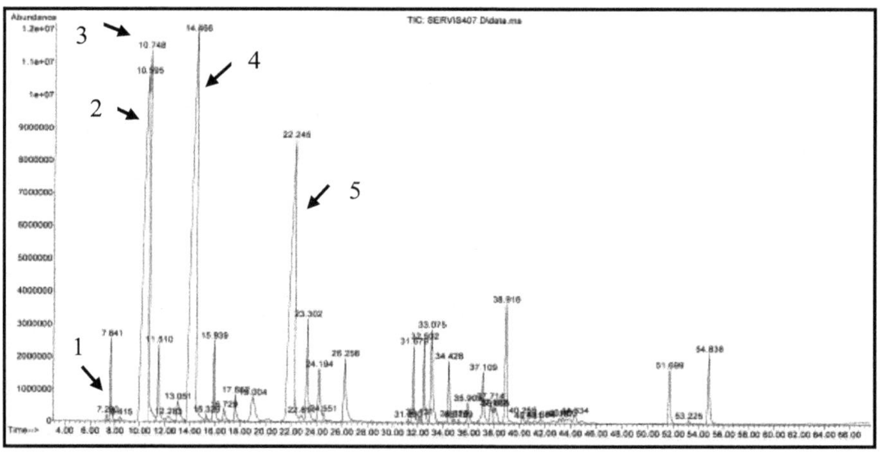

Fig. 8.10 GC-MS chromatogram of volatile compound from kaffir lime peel essential oil at uncontrolled temperature. The major compounds marked as (1) α-pinene, (2) sabinene, (3) β-pinene, (4) limonene, (5) citronellal

lime peel oils such as α-pinene, sabinene, β-pinene, limonene, and citronellal were detected for each temperature profile. Figure 8.11 present the graphical major compounds in each temperature controlled [1].

As refer to Fig. 8.11, the major constituent of α-pinene at 80, 90, 95 °C and uncontrolled temperature were 2.515, 1.806, 2.207, and 1.255%, accordingly. The sabinene compound that indicate the spiciness taste in the oil contributes the highest concentration among all with 45.594%, 36.394%, 37.481%, and 27.498% at 80 °C,

Table 8.3 Compounds of essential oils obtained from kaffir lime peel

Relative peak area (%)						
No.	Compounds	RETENTION INDEX (RI)	80 °C	90 °C	95 °C	Without controlled
1	α-Thujene	0924	0.27	0.18	0.22	0.11
2	**α-Pinene**	**0932**	**2.515**	**1.806**	**2.207**	**1.255**
3	**Sabinene**	**0969**	**45.594**	**36.394**	**37.481**	**27.498**
4	**β-Pinene**	**0974**	**8.974**	**8.597**	**8.714**	**7.147**
5	Mycrene	0988	1.786	1.655	1.656	1.547
6	δ-2-Carene	1001	0.109	0.062	0.086	0.070
7	**Limonene**	**1024**	**28.649**	**32.455**	**32.232**	**28.722**
8	γ-Terpinene	1054	0.310	0.498	0.705	0.988
9	Terpinolene	1086	–	0.152	0.205	0.300
10	Trans-sabinene hydrate	1098	0.146	0.136	–	0.273
11	Linalool	1095	0.541	0.810	0.802	0.921
12	**Citronellal**	**1148**	**8.293**	**10.775**	**9.647**	**17.487**
13	Terpinene-4-ol	1174	0.471	1.120	1.452	2.028
14	α-Terpineol	1186	0.350	0.583	0.668	1.242
15	Citronellol	1223	0.429	1.784	1.012	1.809
16	α-Copaene	1374	0.345	0.494	0.400	0.953
17	E-Caryophyllene	1417	0.280	0.313	0.258	0.725
18	Citronellyl propanoate	1444	0.185	0.391	0.315	0.909
19	Cis muurola-3, 5-diene	1448	0.212	–	0.380	1.156
20	α-Humulene	1452	–	0.097	–	0.242
21	Geranyl propanoate	1476	–	–	–	0.810
22	Germacrene D	1484	0.091	0.430	0.162	0.517
23	β-Selinene	1489	–	0.032	–	–
24	Bicyclogermacrene	1500	–	0.073	–	0.256
25	α-Muurolene	1500	–	0.045	–	0.122
26	δ-Cadinene	1522	–	0.424	–	–
		Total	19	24	19	24

90 °C, 95 °C, and uncontrolled temperature, respectively. The third major compound is β-pinene which provide distinctive aroma where 8.974, 8.597, 8.714, and 7.147% were identified in respective temperature. Limonene that gives distinctive aroma of oranges provide 28.649, 32.455, 32.232, and 28.722% concentration at 80, 90, 95 °C, and uncontrolled temperature, accordingly. Meanwhile, for citronellal compound which indicates the micro and insect repellent activity present 8.293, 10.775, 9.647,

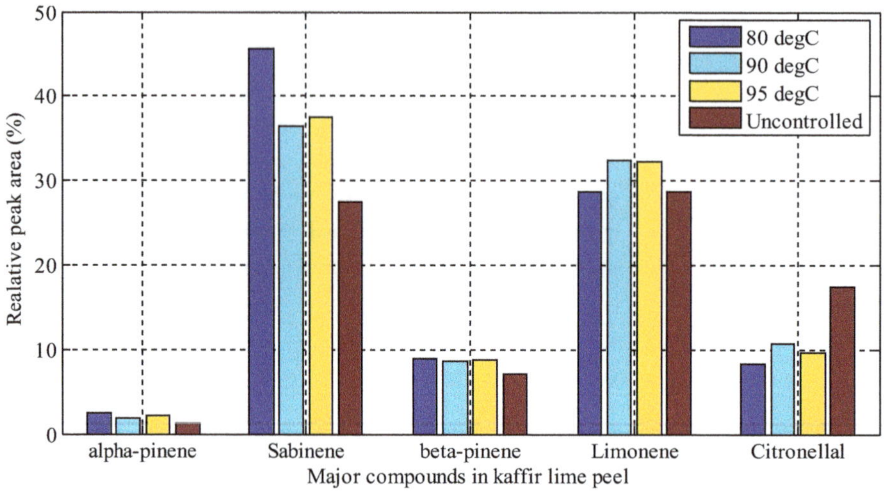

Fig. 8.11 Major compounds in kaffir lime peel

and 17.487% concentration. What is interesting in this data is that it is obvious for compound α-pinene, sabinene, β-pinene, and limonene where the controlled temperature can extract more chemical compounds than uncontrolled temperature. The significant different between these two conditions is due to oil experience thermal degradation which consequently losing some significant compounds in uncontrolled temperature. In contrast to earlier findings, citronellal was notably high in the oils collected from uncontrolled temperature as compared with the controlled temperature. However, high concentration of citronellal will produce rancid. As notice in Fig. 8.12, the color of the oils varies from light to brown as temperature increase [1]. This finding supports previous study where different temperature will give an effect to the number of compounds extracted and quantity of essential oil.

Fig. 8.12 Pure essential oil from kaffir lime peels

Reference

1. Yusoff, Z.M.: Self-tuning Fuzzy PID Control of Hydro-diffusion Essential Oil Extraction System. Universiti Teknologi MARA, Malaysia (2014)

Chapter 9
Final Insights and Future Directions

Abstract This chapter synthesizes the major findings and contributions of the study on intelligent control of hydro-diffusion steam distillation. Addressing the nonlinear dynamics and time-delay challenges inherent in conventional distillation systems, the research employed ARX-based system identification to develop reliable dynamic models for controller design. Comparative simulation and real-time validation demonstrated the limitations of classical PID control and the superior performance of hybrid fuzzy-PID (HFPID) and self-tuning fuzzy-PID (STFPID) strategies. Among them, STFPID-5 consistently achieved the fastest response, lowest error, and strongest disturbance rejection. Beyond engineering performance, the study established a direct link between temperature regulation and chemical quality. Controlled extractions at 90 °C and 95 °C preserved key volatile compounds, producing balanced aroma profiles, whereas uncontrolled boiling at 100 °C, despite higher yield, resulted in chemical degradation. By integrating process control with GC–MS analysis, this work demonstrates that intelligent control enhances both operational efficiency and product integrity. The chapter concludes by outlining future directions toward multivariable control, AI integration, and Industry 4.0-based smart extraction systems.

9.1 Overview

Essential oil extraction has long been a field that combines traditional practices with modern engineering, yet many challenges persist. Conventional hydro-diffusion steam distillation, although effective in principle, suffers from intrinsic limitations such as nonlinear dynamics, significant time delays, and poor controllability of critical variables such as steam temperature. These constraints directly affect the efficiency of oil recovery as well as the chemical profile of the extracted products. In the context of kaffir lime essential oil, which is widely used for its fragrance, medicinal properties, and applications in food and cosmetic industries, maintaining both yield and quality is paramount. However, existing methods often force a compromise

Z. Mohd Yusoff et al., *Fuzzy Logic, Smart Control, and the Future of Essential Oils*, Studies in Fuzziness and Soft Computing 442,
https://doi.org/10.1007/978-981-92-2084-7_9

between maximum yield and chemical stability, as excessive heating can degrade delicate volatile compounds even if the oil recovery is high.

This book addressed these challenges by systematically investigating how intelligent control systems can improve the performance of hydro-diffusion steam distillation. The research journey began with modeling and system identification using ARX techniques. This step was critical because accurate models form the foundation for designing effective controllers. The study then advanced into simulation analysis, where classical PID controllers were compared against advanced intelligent controllers such as the hybrid fuzzy-PID (HFPID) and the self-tuning fuzzy-PID (STFPID). These simulations provided the first insight into how intelligent approaches could mitigate the sluggish response and poor adaptability of conventional PID.

The next stage involved real-time implementation and validation of these controllers in a physical hydro-diffusion distillation setup. This phase was crucial to demonstrate that the advantages observed in simulations were not theoretical artifacts but could be replicated in practice under real disturbances and nonlinearities. The findings confirmed that intelligent controllers, particularly the STFPID with five membership functions (STFPID-5), provided faster response, greater robustness, and higher reliability compared to conventional PID.

Finally, the study extended its validation to the chemical analysis of extracted kaffir lime oil, linking engineering performance directly with product outcomes. Using controlled heating regimes, it was shown that operating at 90 °C and 95 °C preserved the balance of key volatile compounds such as sabinene, limonene, β-pinene, and α-pinene. In contrast, uncontrolled extraction at 100 °C, although yielding the highest oil percentage, resulted in chemical degradation, dominated by excessive citronellal that compromised aroma quality. This integration of process control with GC-MS analysis represents a significant multidisciplinary contribution, demonstrating that intelligent control is not only about numbers and performance indices but also about tangible improvements in product quality.

Placing this study within the broader global context highlights its relevance. The essential oil industry is expanding rapidly, driven by increasing consumer demand for natural products in pharmaceuticals, cosmetics, aromatherapy, and food additives. At the same time, there is a strong push towards sustainable and efficient bioprocessing methods, aligned with the principles of Industry 4.0 and green chemistry. Intelligent process control fits squarely within these trends, offering a pathway to higher yields, lower energy consumption, and products of superior quality. This study therefore contributes not only to academic literature but also to industrial innovation and societal benefit, reinforcing the idea that engineering solutions must align with environmental responsibility and consumer expectations.

9.2 Synthesis of Key Findings

The findings of this work can be organized into three major stages:

- Simulation analysis.
- Real-time validation.
- Experimental extraction with chemical profiling.

Each stage provided distinct yet interconnected insights that, when combined, form a coherent picture of how intelligent control systems can significantly improve the performance of hydro-diffusion steam distillation.

The first stage, presented in Chap. 6, dealt with simulation-based performance analysis of conventional PID, HFPID, and STFPID controllers. Simulations are a necessary starting point for process control studies because they allow for systematic comparisons under controlled conditions without the interference of experimental noise. The results clearly demonstrated the limitations of the classical PID controller. Although stable and widely used, PID showed sluggish dynamics, with longer rise and settling times. While overshoot was effectively eliminated, the response lacked agility, making it unsuitable for processes characterized by significant time delays.

By contrast, the HFPID controllers introduced nonlinear adaptability by embedding fuzzy rules into the PID framework. This enhancement allowed the controller to respond more efficiently to changing conditions. Among the variants, HFPID with seven membership functions (HFPID-7) provided the most balanced transient response, showing shorter rise and settling times compared to both PID and HFPID-3. The STFPID controllers extended this adaptability by dynamically adjusting their parameters in real time. STFPID-5 emerged as the best performer across all tested indices, achieving the lowest RMSE and negligible steady-state error. These findings, summarized in Table 9.1, underscore the value of fuzzy-based enhancements to traditional control methods. The simulation evidence was which compared RMSE values across controllers and clearly highlighted the superiority of STFPID.

The second stage, described in Chap. 7, involved real-time implementation and validation of these controllers. This step was crucial in confirming that the advantages observed in simulation were not confined to theoretical models but could withstand the complexities of actual process conditions. Real-time experiments revealed that the PID controller, while stable, was slow to recover from disturbances, making it a weak performer in dynamic environments. HFPID controllers, particularly HFPID-7, demonstrated faster recovery and greater robustness, aligning with the simulation results. However, it was the STFPID controllers, and especially STFPID-5,

Table 9.1 Comparative simulation performance of controllers

Controller	Rise time	Settling time	RMSE	Steady-state error	Disturbance rejection
PID	High	Long	0.85	Small	Weak
HFPID-3	Medium	Moderate	0.55	Very small	Moderate
HFPID-7	Short	Faster	0.40	Negligible	Strong
STFPID-3	Very short	Fast	0.30	Negligible	Strong
STFPID-5	Shortest	Fastest	0.20	Negligible	Strongest

Table 9.2 Comparative real-time validation performance of controllers

Controller	Recovery time	RMSE	Disturbance rejection
PID	Slow	0.95	Weak
HFPID-3	Moderate	0.60	Moderate
HFPID-7	Fast	0.42	Strong
STFPID-3	Very fast	0.28	Strong
STFPID-5	Fastest	0.18	Strongest

that consistently achieved the best outcomes. With rapid setpoint tracking, excellent disturbance rejection, and the lowest RMSE among all controllers, STFPID-5 proved to be the most reliable and adaptable under practical operating conditions. These results, shown in Table 9.2, confirmed that intelligent controllers can indeed translate theoretical benefits into real-world performance improvements.

The third and perhaps most consequential stage of the study, presented in Chap. 8, linked controller performance to product outcomes in kaffir lime oil extraction. Here, the hydro-diffusion steam distillation system was operated at controlled temperatures of 80 °C, 90 °C, and 95 °C using intelligent controllers, as well as at an uncontrolled boiling point of 100 °C. The yield results revealed a clear trend: yield increased with temperature, peaking at 3.11% under uncontrolled conditions. However, uncontrolled extraction compromised chemical integrity. GC-MS analysis showed that the oil extracted at 100 °C was dominated by citronellal at 17.5%, with significant losses of sabinene, limonene, and α-pinene. This imbalance produced a rancid aroma, undermining product quality despite the higher yield.

In contrast, controlled extractions at 90 °C and 95 °C produced more balanced chemical profiles. At 90 °C, the yield was 2.19%, and the oil contained strong concentrations of sabinene (36.4%) and limonene (32.5%), producing a desirable aroma profile. At 95 °C, yield was slightly higher at 2.94%, and although citronellal increased modestly to 9.6%, the balance among compounds was maintained. Extraction at 80 °C produced the lowest yield (0.70%) and was therefore economically unviable, despite acceptable compound preservation. These findings are summarized in Table 9.3.

A more detailed look at the GC-MS results provides further justification for the superiority of controlled heating. Table 9.4 compares the percentage composition of

Table 9.3 Extraction yield of kaffir lime oil under different temperatures

Temperature (°C)	Yield (%)	Aroma quality	Key observations
80	0.70	Good	Yield too low, not viable
90	2.19	Excellent	Best aroma profile, balanced compounds
95	2.94	Very good	Slightly higher yield, good compound balance
100	3.11	Poor	High yield but degraded aroma, excessive citronellal

the five major compounds across temperature conditions. The controlled extractions at 90 °C and 95 °C preserved sabinene and limonene in the range of 36–37% and 32%, respectively, compared to only 27% sabinene and 28% limonene under uncontrolled boiling. This confirms that while uncontrolled extraction maximizes yield, it does so at the expense of chemical quality. By contrast, intelligent control strategies balance efficiency with integrity, producing oils of higher commercial and sensory value.

Taken together, the findings from simulation, real-time validation, and extraction experiments form a consistent and compelling narrative. The superiority of intelligent controllers was demonstrated in both theory and practice, and their benefits extended beyond process control metrics to the actual product. The integration of results across these stages is summarized in Table 9.5, which presents a holistic view of controller performance from simulation to chemical analysis. It is evident that STFPID-5 stands out as the best overall performer, excelling in both control performance and product outcomes.

Intelligent control is not merely an academic exercise in optimizing response curves but a practical necessity for improving both efficiency and quality in essential oil extraction. By ensuring stable, adaptable, and precise temperature regulation, intelligent controllers directly influence the yield and chemical profile of extracted oils. In doing so, they bridge the gap between engineering design and consumer product quality, offering a multidisciplinary contribution that has implications for academia, industry, and society alike.

Table 9.4 Percentage composition of major compounds at different temperatures

Compound, (%)	80 °C	90 °C	95 °C	100 °C
α-Pinene	12.5	14.8	15.0	11.2
Sabinene	35.7	36.4	37.1	27.0
β-Pinene	6.2	7.1	7.0	5.5
Limonene	31.5	32.5	32.0	28.0
Citronellal	6.5	7.2	9.6	17.5

Table 9.5 Integrated summary of controller and extraction performance

Aspect	Best performer	Key notes
Simulation	STFPID-5	Lowest RMSE, fastest response
Real-time validation	STFPID-5	Fastest recovery, strongest robustness
Extraction yield	Controlled 95 °C	High yield without excessive degradation
Compound quality	Controlled 90 °C	Preserved sabinene and limonene, balanced profile

9.3 Contributions and Novelty

The findings presented in this book has made several significant contributions that extend across technical, practical, and multidisciplinary dimensions. At its core, the study addressed a persistent challenge in essential oil extraction, the difficulty of controlling nonlinear and delay-dominated processes using conventional methods. By designing, simulating, validating, and experimentally applying intelligent controllers, this work demonstrated not only theoretical advancements in control system design but also tangible improvements in extraction yield and product quality.

From a technical perspective, one of the key contributions lies in the successful application of ARX modeling for system identification of a hydro-diffusion steam distillation process. While ARX models are well known in control literature, their use in capturing the dynamics of essential oil extraction systems is relatively novel. This modeling approach provided a reliable mathematical foundation for controller design, enabling accurate prediction of system behavior in both simulation and real-time conditions. In doing so, the study bridges the gap between abstract mathematical modeling and its practical application in a traditionally artisanal process.

Another major contribution is the design and implementation of fuzzy-based hybrid controllers. The HFPID controllers introduced nonlinear adaptability that overcame the sluggishness of classical PID controllers. More importantly, the STFPID controllers represented a substantial leap forward by adjusting control parameters dynamically in response to system changes. Among these, the STFPID-5 variant consistently outperformed all others across simulation and real-time tests, providing the lowest RMSE, fastest recovery, and highest robustness. This confirms the novelty of combining self-tuning mechanisms with fuzzy inference to address the inherent nonlinearities of bioprocesses. The results in Tables 9.1 and 9.2 support this claim by showing clear improvements in performance indices when moving from PID to intelligent controllers.

On the practical side, the research demonstrated that controller design is not an isolated engineering exercise but one with direct implications for product outcomes. Chap. 8 showed that temperature control strategies influenced not only the quantity but also the chemical quality of kaffir lime essential oil. At controlled temperatures of 90 $°C$ and 95 $°C$, the extraction process preserved the balance of volatile compounds, producing oils with superior aroma profiles. By contrast, uncontrolled boiling at 100 $°C$ yielded slightly more oil but degraded its composition, resulting in excessive citronellal that compromised aroma. The findings summarized in Tables 9.3 and 9.4 highlight that process control decisions directly shape consumer product quality, a perspective rarely emphasized in control systems research.

A further contribution lies in the novel integration of engineering and chemical sciences. Previous studies on essential oil extraction have often focused on either process optimization or chemical profiling in isolation. This work, however, brought the two together, establishing a direct link between controller performance and chemical composition outcomes. This integration is not only academically novel but also practically relevant, as it allows industries to optimize for both efficiency and quality simultaneously. The integrated summary provided in Table 9.5 illustrates this contribution by unifying results across simulation, real-time validation, and extraction experiments, showing that the same controller (STFPID-5) that excelled in engineering terms also contributed to superior product outcomes.

Finally, this study contributes to multidisciplinary knowledge creation by demonstrating that intelligent control frameworks can be successfully adapted for traditional natural product industries. By embedding fuzzy logic and self-tuning adaptability into a distillation process, the research aligns with global trends in Industry 4.0, where smart, adaptive, and data-driven technologies are transforming manufacturing. The novelty here lies not only in the controllers themselves but also in their application to an industry that typically relies on conventional or manual practices. This opens new opportunities for applying advanced control theories to domains such as food processing, herbal medicine production, and bioproduct industries, where maintaining quality and sustainability is as important as achieving efficiency.

9.4 Recommendations for Future Research

While this study has provided valuable insights into the application of intelligent controllers for essential oil extraction, it also opens new avenues for further exploration. The findings demonstrate the potential of fuzzy-based and self-tuning control strategies, yet they also highlight the limitations of the current research scope. Future studies can build upon this foundation by expanding the scope of variables, methods, and applications to achieve even greater improvements in efficiency, quality, and sustainability.

One of the most immediate directions for future research is the expansion of control variables beyond temperature. This study focused exclusively on steam temperature, which indeed plays a central role in essential oil extraction. However, other process parameters, such as steam pressure, flow rate, and feedstock loading, also influence yield and quality. Developing multivariable control systems that regulate several parameters simultaneously would provide a more comprehensive solution. Such an approach would allow finer control over the extraction process, minimizing the risk of degradation while maximizing recovery. The integration of multivariable fuzzy or model predictive control systems could therefore represent a significant advancement in process regulation.

Another promising direction is the adoption of advanced artificial intelligence techniques. While fuzzy logic provides a strong foundation for handling nonlinearities, emerging AI tools such as reinforcement learning, neural networks, and deep learning offer even greater adaptability. Reinforcement learning could allow controllers to continuously improve their performance by learning from process outcomes in real time. Neural networks could be used for predictive modeling, enabling proactive adjustments before deviations occur. By combining these approaches with fuzzy logic, future systems could achieve a higher level of intelligence, adaptability, and robustness.

Scaling up the research from laboratory-scale experiments to industrial-scale applications is also essential. While the laboratory results are promising, large-scale operations present additional challenges such as longer process times, variability in raw material quality, and higher energy demands. Pilot-scale or industrial-scale validation would provide critical insights into the robustness, cost-effectiveness, and practicality of intelligent controllers in commercial settings. Such work would also help bridge the gap between academic research and industrial adoption, encouraging companies to invest in digitalization and automation of their extraction processes.

In addition to scaling up, future studies should explore the generalization of the methodology to other plant species and extraction systems. Kaffir lime was chosen for this research due to its high commercial value and aromatic importance, but the principles developed here are equally applicable to other essential oils such as lemongrass, citronella, agarwood, and peppermint. By testing the framework across different raw materials, researchers can assess the universality of the intelligent control approach and tailor optimization strategies to specific chemical compositions. This would further enhance the contribution of intelligent process control to the broader essential oil industry.

Future research should also extend beyond chemical profiling to include bioactivity testing and sensory evaluation. While this study analyzed chemical composition using GC-MS, it did not directly assess the biological properties or consumer perception of the oils. Antimicrobial, antioxidant, and anti-inflammatory assays could provide insights into the therapeutic potential of the oils, while sensory panels could evaluate aroma quality from a consumer perspective. By linking control strategies to both bioactivity and sensory outcomes, future studies could provide a more holistic assessment of how process optimization translates into product functionality and marketability.

Finally, the integration of intelligent distillation systems with Industry 4.0 technologies represents a forward-looking recommendation. The combination of IoT-enabled sensors, cloud-based data storage, and AI-driven analytics would allow real-time monitoring, adaptive optimization, and remote process management. Such a framework could make essential oil extraction more efficient, transparent, and scalable, particularly in decentralized production environments.

In summary, the recommendations for future research can be grouped into five key areas: expanding to multivariable control, adopting advanced AI methods, scaling up to industrial applications, testing across different plant species, and integrating chemical profiling with bioactivity and sensory studies. Collectively, these directions point toward a future where essential oil extraction is not only more efficient and consistent but also more sustainable, consumer-oriented, and digitally integrated. Such advancements would ensure that the essential oil industry remains competitive, resilient, and aligned with global trends in smart manufacturing and natural product innovation.

Index

Z. Mohd Yusoff et al., *Fuzzy Logic, Smart Control, and the Future of Essential Oils*, Studies in Fuzziness and Soft Computing 442,
https://doi.org/10.1007/978-981-92-2084-7